Metallurgical Advances in Coatings and Corrosion

This book offers a unique narrative and provides insights into the innovative solutions pertaining to corrosion in industrial settings. From safeguarding maritime structures to enhancing the performance of diesel engine bearings in harsh environments, the anthology explores various facets of corrosion mitigation and materials enhancement. Chapters delve into topics such as protective coatings, electrodeposition techniques, alloying effects, and electrochemical dynamics crucial for understanding corrosion mechanisms and resistance.

Features:

- Offers a comprehensive exploration of metallurgy and corrosion research, providing insights into various aspects of materials science and engineering advancements.
- Covers diverse topics such as protective coatings, electrodeposition, alloying effects, and electrochemistry.
- Each chapter presents real-world applications and solutions to corrosion-related challenges, offering actionable insights that can be applied in industrial settings.
- Focuses on metallurgical advancements and corrosion mitigation strategies.
- Reviews sustainable production, health, and environmental impacts.

This book is aimed at graduate students and researchers in materials and metallurgical engineering.

Metallurgical Advances in Coatings and Corrosion

Ojo S.I. Fayomi

CRC Press is an imprint of the
Taylor & Francis Group, an **informa** business

Designed cover image: www.shutterstock.com

First edition published 2025
by CRC Press
2385 NW Executive Center Drive, Suite 320, Boca Raton FL 33431

and by CRC Press
4 Park Square, Milton Park, Abingdon, Oxon, OX14 4RN

CRC Press is an imprint of Taylor & Francis Group, LLC

ISBN: 9781032891293 (hbk)
ISBN: 9781032912356 (pbk)
ISBN: 9781003562078 (ebk)

DOI: 10.1201/9781003562078

Typeset in Times
by Newgen Publishing UK

Contents

Preface

"Metallurgical Advances in Coatings and Corrosion" is a collection that goes beyond the normal limitations of metallurgy and corrosion research. This anthology is a testimony to the never-ending search for knowledge, bringing together research reports that reveal the intricate fabric of materials science and engineering advancements. Each numbered chapter tells a distinct story, a portal to a world where problems become solutions and the never-ending battle against corrosion reshapes industrial landscapes. Chapter 1, "Safeguarding the Integrity of Maritime Structures", leads an expedition through protective coatings and advanced deposition methods, fortifying metals against degradation in marine environments. Chapter 2, "Electrodeposition: Effects of Zinc Multifaceted Barrier Protection in Sustaining Engineering Structure", showcases the adaptability of electrodeposition across diverse applications, addressing challenges impacting critical industries. Chapter 3, "Diesel Engine Bearings in Harsh Environments", underscores the strategic role of coatings in enhancing bearing performance amid extreme conditions. Chapter 4, "Corrosion: A Science of Degradation", unravels the complexities of corrosion forms and mitigation strategies, emphasizing the fusion of science and invention in protective coatings. Chapter 5, "Deposition of Binary and Quaternary Alloys on Steel for Maximal Performance Use", examines alloying effects and explores novel deposition methods to enhance mild steel's qualities. Chapter 6, "Corrosion Multifaceted Impact on Mild Steel", uncovers the intricate interactions influencing mild steel's vulnerability to corrosion, advocating for a multifaceted approach to mitigation. Chapter 7, "Aluminium Composite Development for Sustainable Production", showcases advancements and future directions for aluminium alloys across various industries. Chapter 8, "Electrochemistry of Corrosion", delves into electrochemical nuances crucial in understanding corrosion dynamics and resistance mechanisms. Chapter 9, "High Entropy Alloys: The Study on Aluminium Cooking Ware", addresses concerns surrounding aluminium deterioration in cookware, spotlighting health and environmental impacts, and proposing solutions through collaborative efforts.

Professor Ojo S.I. Fayomi
Department of Mechanical Engineering Science
Auckland Park, Kingsway Campus Johannesburg,
South Africa Department of Mechanical Engineering
Bells University of Technology Ota, Ogun,
Nigeria February 2024

About the Author

Ojo S.I. Fayomi is a Full Professor, current Executive Dean of the College of Engineering, and Chairman, Corporate Affairs Unit, Office of the Vice Chancellor, Bells University of Technology, Ota, Nigeria. He is an internationally renowned researcher, teacher, and life coach. He is an established proven academia and global researcher in the area of Mechanical Metallurgy of Materials and Production Technology with a focus on marine, automotive, aerospace, manufacturing, and biomedical applications. He obtained a National Diploma (Distinction) from the Polytechnic Birnin Kebbi, Kebbi State, Nigeria, a Higher National Diploma in Metallurgical and Materials Engineering from Federal Polytechnic Idah, Kogi State with (Upper credit), a Bachelor of Engineering (Honours) in Mechanical Engineering from Bells University of Technology. He bagged Master's and Doctoral degrees in Metallurgical Engineering with Distinction (Cum Laude) from the Department of Chemical, Metallurgical and Materials Engineering, Tshwane University of Technology, Pretoria, South Africa. He has a postgraduate diploma in environmental and natural resource management from Olabisi Onabanjo University, Ago-Iwoye.

Professor Ojo S.I. Fayomi has been instructing in the University and research environment since 2007. He rose to be a distinguished exceptional Professor with a massive contribution to new materials development for multifaceted applications. He is a visiting scholar at several Universities locally and internationally. He is a visiting research director at Surface Engineering Research Centre, a Research and Innovation Associate with Tshwane University of Technology; Pretoria (2018–2023), and presently an extraordinary professor with Tshwane University of Technology; Pretoria (2023 till date). Due to his vast expertise, the University of Johannesburg engages him as a Senior Research Associate in the Faculty of Engineering. He has to his credit over (500) indexed international and local peer-reviewed journals with substantial impact factors, broad readership, and citations in Scopus and Web of Science. Professor Fayomi is an External Examiner to several Nigerian universities at both undergraduate and postgraduate levels, and some foreign universities. He has supervised over 150 international research/industrial projects, and graduated 15 doctoral degree holders and 40 Master's degree students apart from the ongoing postgraduate supervision.

Professor Ojo S.I. Fayomi is a sought-after inspirational public speaker and lecturer to many industries, universities, and institutes. Professor Fayomi is known to be a reviewer of highly rated Scientific Journal and Conference publishers in related fields such as Elsevier, Springer, Taylor and Francis, Wiley, and others. He is also an examiner to the National Research Foundation, South Africa, Portugal, and Technical Chair of one of the most soughed largest researcher conference gatherings in Africa, "ICESW". He is a Co-chairman of Standards and Dredgers with the Standard

Organisation of Nigeria. He is a member of several engineering bodies such as the Council for the Regulation of Engineering in Nigerian (COREN), Member of Nigeria Society of Engineers (NSE), a Registered Academic Member of the Water Institute of Southern Africa (AMWISA), Member Nigeria Metallurgical Society (NMS), and Member Nigeria Association of Technologist in Engineering.

He is an achieved researcher and lecturer, who has harvested several awards internationally. Among others is the Scientific Achievement by South African Association for Advancement in Science Medal Award (A2S3 Medal for Original Research). Fayomi is a Postdoc Research Scholar and an Awardee of the Institutional Best Postdoctoral Fellow for Teaching and Innovations at Tshwane University of Technology. He also emerged as the best doctoral degree award student at Tshwane University of Technology. He is well-reputed nationally and internationally in the areas of Mechanical Materials Mechanics. He was named the Outstanding African Researcher, in 2019, and the Most Innovative Scholar in Composite Deposition. In 2021, Professor Ojo S.I. Fayomi was listed among the top 50 in Africa out of the top 500 researchers, by scholarly output, over the period 2015–2020 by Scival. He has a patent to his credit with the South Africa Ministry of Education and Cooperates affair.

Professor Ojo S.I. Fayomi is a passionate, purpose-driven, self-motivated, and result-oriented natural leader. He is a multi-talented person, perfectionist, and president of Ignite World—an organization founded by him to help the needy, youth, and family discover self. Professor Ojo S.I. Fayomi is the founder of a monthly publication called DISCOVERY, a network of vibrant conference and inspiration presentations. He has published many weekly motivational and transformational materials through hard work, commitment, and dedication to the vision. He has trained hundreds to thousands of people through Ignite World motivational summits, seminars, workshops, and conferences. He presently serves as a mentor to several people in Nigeria and around the world. As an international conference speaker, he has spoken to audiences in Belgium, Mauritius, the United States, Grease, the United Kingdom, South Africa, and Nigeria.

Today, Professor Ojo S.I. Fayomi is a multi-gifted inspirational speaker who addresses critical issues that affect every aspect of human concepts, professional practices, and social and inspirational empowerment. Fayomi is happily married to his wife Dr. Gloria Uche Fayomi, and they are blessed with two children Isaac Jnr and Triumph.

1 Safeguarding the Integrity of Maritime Structure

Overview of Coating Techniques

1.1 INTRODUCTION

The application of mild steel for structural outcomes, fabrications, and other fundamental engineering and mechanical applications cannot be overstated (Javidan et al., 2016). The structure of mild steel is a composition of several metallic materials assembled by crystallization. The output from the crystallization results in a mild steel and it assists in an added dimension in manufacturing industries to serve as a raw material for the construction of ships (Lehmhus et al., 2013). In its various applications, the vital features consist of marine grade mild steel, the material must be able to resist the corrosive effects that are common in a water environment (Nevshupa et al., 2018; Jumbo et al., 2020), and mild steel of various grades, but the Q235 grade of mild is widely used for ship-building application due to the fact that the Q235 not only possess toughness, held ability, and excellent plasticity but also retains strength while providing excellent cold-bending performance. The recognition and use of mild steel in the maritime industries essentially for ship-building is a reality that calls for awareness. This is because marine vessels predominantly sail in a corrosive environment such as seawater, which is extremely intense with brackish solution (Guedes Soares & Garbatov, 2011). The huge percentage of the saltwater solution established in oceans, seas, and rivers is the primary cause of corrosion of marine vessels constructed with mild steel (Yang et al., 2019). From this perspective, marine diesel components like the hulls, the bulkheads, and other crucial parts have deteriorated from this hazard. It has jeopardized the maritime industry as the vessels' hulls constantly immersed in water can rust and sometimes lead to the ingress of water into the cargo holds posing a potential risk to the ship. Conversely, with these remarkable vital limitations of the maritime vessel-based mild steel components, a bit or no statistics are documented to increase the quality of this engineering material for utilization in the maritime industry. However, the evolution of new research topics has been centred towards enhancing the corrosive resistance of mild steel by using a suitable surface protection product for the exposed parts of the ship.

DOI: 10.1201/9781003562078-1

1.2 PROTECTION OF STEEL

To overcome the problems that come with rust and other corrosion forms that can lead to safety issues, economic deficits, and probably destroy the integrity of the machine components, five methods for corrosion prevention have been established. These techniques involve the use of barrier coatings, hot-dip galvanization, alloyed steel (stainless), cathodic protection, and EonCoat methods (Toklu et al., 2018; Yu et al., 2020; Y. Li et al., 2019).

1.2.1 Barrier Coating

This stands as one of the simplest and cost-effective techniques to prevent corrosion. This involves the use of substances like paint, plastic, or powder. Powders including epoxy, nylon, and urethane stuck to the metal substrate to form a thin film. The plastic and wax are normally applied onto the metal substrates. Paint acts as a coating to prevent the metal substrate from the electrolytic charge that comes from the corrosive compounds. The biggest drawback with coating is that it often needs to be stripped and reapplied (Zubielewicz et al., 2021).

1.2.2 Hot-Dip Galvanization

This technique involves submerging steel into molten zinc. This process is currently the main metallic coating. Getting a superior surface appearance in galvanized steels is directly proportional to knowing the phenomena involved in the coating techniques. The iron in the steel reacts with zinc to create a firmly bonded alloy coating, which acts as a prevention. This technique has been used for more than 250 years and has been applied for corrosion protection of materials like artistic sculpture and playground equipment (Vieira et al., 2022).

1.2.3 Cathodic Protection

Cathodic protection is a mechanism of preventing corrosion of a metal surface by making it the cathode of an electrical circuit. It is used to protect steel pipelines, storage tanks, steel pipes, ships, offshore oil platforms, and gas and water pipes (Loto & Popoola, 2011; Loto et al., 2019). Cathodic protection prevents corrosion by electrolytic method. To protect against corrosion, the active areas on the metal surface are turned into passive areas by offering electrons (ions) from another origin, predominantly with galvanic anodes attached to the surface. Surface applied as anode are normally aluminium, magnesium, or zinc. Cathodic protection is ultimately efficacious, and anodes get used up and need to be checked and/or replaced often.

1.2.4 Metal Alloying

This is one of the most efficacious corrosion prevention techniques today, adding the attribute of various metals to give more hardness and resistance to the revealed product. Corrosion-resistant nickel, for instance, combined with oxidation-resistant chromium forms an alloy that can be applied in an oxidized and reduced chemical environment. But alloys of steel such as stainless steel have a chromium component of 10.5% (Bolli et al., 2020; Y. Li et al., 2019).

1.2.5 EonCoat

This is one of the newest techniques to protect substances from corrosion. EonCoat is a very cost-effective technique; in other words, it is maintenance-free and is an easily used solution that prevents the life of the material from corroding. This technique works perfectly with the combination of the methods (Liu et al., 2022).

1.3 IMPACT OF METAL COATING ON SURFACES AND STRUCTURES

The potential impact of coatings on metal surfaces and structures is to withstand corrosion and enhance other attributes such as the mechanical properties of the material that has been generally attributed to the development of a hard-thin layer connecting the material and coating interface, which gives a reinforcing for the material (Cui et al., 2018; Jabbar et al., 2017). However, metal corrosion is a degradation process that occurs under specific circumstances. The most common type of corrosion happens when metals combine with moisture and oxygen to form several corrosive products. Iron, for instance, combines with water and air to produce iron III oxide or rust (Mai et al., 2022). The logic behind metal coating, therefore, is to form an inert barrier around the metallic object being protective to prevent it from reacting with oxygen and moisture.

1.4 TYPES OF METAL COATINGS

Anodizing: Anodizing is a type of metal coating applied to promote the production of a protective oxide layer on the metal surface (Rashid & Khadom, 2022). The resulting oxide layer forms more rapidly and is normally thicker than if it was formed naturally. While many non-ferrous metals can be anodized, aluminium responds most effectively to this process.

Galvanizing: This type of metal coating involves dissolving the metal in a molten zinc bath. Once taken from the immersed solution, the coated metal combines with air and carbon dioxide in the atmosphere to create a protective zinc carbonate layer (Park et al., 2022).

Electroplating: This is also referred to as electrodeposition of metals. It involves depositing a thin film layer of one metal on the surface of another metal. During electroplating, both metals are placed in an electrolytic solution. The metal to be coated acts as anode while the coating metal acts as the cathode. An electric current is applied to the electrolytic cell, causing the migration of ions from the cathode to the anode, thus forming the coating (Naveed et al., 2022).

Powder coating: This is a type of metal coating that involves coating a material with a powder-based material. It is an electrostatic process, whereby the coating particle is electrically charged with a polarity. This is contrary to the part to be coated (Xia et al., 2013).

Paint coating: This type of metal coating is predominantly the application of liquid paint. It is the most accessible and cost-effective type of coating. Several paint configurations can be applied depending on the type of metal, the operating environment, and the performance required (Mohanty et al., 2023).

1.5 COMPOSITE ELECTRODEPOSITION

Composite electrodeposition has the capacity to form a wide range of coatings by integrating particles into an electrodeposit. The matrix can be a metal substance or conductive polymer, where the particle can be metallic, polymeric, ceramic, or a combination of spherical, irregular, or layered additions. Composite electrodeposition involves both faradaic and electrophoretic processes. It has also been reported that composite coating formed by electrodeposition possesses vital mechanical, tribological, and electrochemical attributes (Walsh et al., 2020).

1.5.1 Mechanism of Particle Amalgamation

A lot of scholars in the field of metallurgy have tried to examine the dynamics of deposition to unravel the way particles are integrated into metal matrix over the years. Many illustrations in the form of models have been established that involve one or more of the following processes:

- Movement of the charged particles to the cathode by electrophoresis;
- Dissolution of the particles on the surface of the electrode by van der Waals force;
- Mechanical integration of the particles into the layer (Costa et al., 2020).

1.6 EFFECTS OF OPERATING PARAMETERS

Integration of composite nanoparticles into metal coating relies predominantly on hypothetical parameters (Jouyandeh et al., 2019). The outcome and the quality of the produced composite coating depend on the quantity and pattern of particle incorporation in the metal matrix. Such hypothetical parameters are the applied current density, particle characteristics and concentration in the bath, temperature of the bath, electrolytic pH, deposition time, chemical additives, and bath agitation.

1.6.1 Applied Current Density

Current is one of the vital operating parameters, the reason being that the coating characteristics depend on current. This method applies electrical current to quicken the absorption and migration of electrons to be deposited at the cathode. The report revealed that at lesser current densities, stronger and well-refined coatings with optimal concentration of incorporated particles are produced (Fayomi et al., 2016).

1.6.2 Particle Characteristics and Concentration in the Bath

The size, shape, nature, and concentration of the particles applied for the production of composite coating have remarkable effects on the morphologies by many types of integration dynamics of co-deposition and features of the next coatings. The incorporation of metal oxides, carbides, nitrides, borides, etc. has been slated to reveal fine surface morphologies by many researchers (Portela et al., 2020; Mohanty et al., 2019).

1.6.3 Bath Temperature

Temperature plays a higher role in electroless plating than in composite electrodeposition. Conversely, augmenting the bath temperature is revealed to improve reaction rates in chemical operations. Excessive temperatures raise the provision of ions to the cathode and the rate of seeding of particles at the cathode (Wei et al., 2018).

1.6.4 Electrolytic pH

The bath pH has an impact on the actions of the bath since reactions are known to only proceed under definite pH states (Ibrahim & Bakdash, 2020; Unal & Karaham, 2018):

$$2H_2 + 2i = H_2 + OH^- \tag{1.1}$$

$$2H_2O + 2i = H_2 + OH^- \tag{1.2}$$

1.6.5 Time

The extent and quantity of the amalgamated nanoparticles depend predominantly on the deposition duration (Meng et al., 2023; Mazare et al., 2018).

1.6.6 Chemical Additives

Chemical additives execute a huge function in electrodeposition. The chemicals (surfactants and stabilizers) improve the radiance, impel grain treatment, and enhance deposit attributes (Sun et al., 2020).

1.6.7 Agitation of the Bath

The agitation of the bath serves two purposes. The agitation of the bath serves two purposes during electrodeposition: it keeps the nanoparticles in suspension and the movement of the particles to the cathode for co-deposition; it also increases the electrolyte current and permits a high functional current density (Bates et al., 2015; Zhang et al., 2021).

1.7 SHIP HULL COATING

The steel or aluminium hulls of ships and boats need to be prevented from the corrosive effects of water so that they do not corrode and fall apart. More so, these hulls and also non-corroding hulls such as those made of glass-reinforced plastic are prone to accumulating bio-fouling marine plants and animals which attach themselves to ship hulls as they do to any suitable underwater surface for the hulls. This fouling can greatly increase resistance and prevent the ship from gliding smoothly through the water. This in turn results in greater fuel consumption in order to get the vessel from point A to point B and that then causes excessive volumes of emission of greenhouse gases and other atmosphere pollutants. The bio-fouling when it becomes severe can also damage the coating and the hull itself (Nwuzor et al., 2021).

1.8 SUGGESTED COATING SYSTEM FOR DIFFERENT AREAS OF A SHIP

Marine vessels consist of a huge variety of service environments that each require protection from the elements (Barboza et al., 2019). Many coatings systems have been developed to protect every part of a marine vessel, from bow to stem, bridge to hull, and every space in between given below are some of the examples of coating systems that perform well in those environments.

1.8.1 Tank Coating System

Ship is full of tanks, each with a unique function in terms of tank coating systems, one size does not fit all. Consider the following different types of tanks (Zhuang et al., 2019):

- Liquid cargo tanks – the coatings in tanks responsible for carrying crude oil, chemicals, or other liquids must protect the underlying steel from corrosion and must be inert towards the products they contact.
- Fresh water tanks – the coatings in these spaces must protect steel and not impact off-tastes to potable water (Kirchgeorg et al., 2018).
- Ballast tanks – these areas feature varying levels of saltwater and warm temperatures making them ideal for corrosion to occur. The safety of life at sea (SOLAS) convention contains provisions on corrosion and prevention in seawater ballast tanks (Abbas et al., 2020).

1.8.2 Engine and Equipment Spaces

Colour matters in interior spaces such as engine rooms, work areas, storage spaces, and common areas. Lighter colours help reflect light and improve visibility. Cleanliness is also important in these spaces, so coatings that are easy to clean are recommended.

Bulkheads and overheads are best protected by the following:

- Epoxy polyamides;
- Surface-tolerant epoxies;
- Long-oil alkyd primers with alkyd undercoats and topcoats.

1.8.3 Interior and Exterior Decks

Decks are best protected by epoxy polyamides or surface-tolerant epoxies. If deck surfaces will sustain abuse from cargo or equipment, a topcoat of glass fibre-reinforced epoxy is recommended. Interior and exterior surfaces that will see high traffic or moisture should have non-skid additives as part of any coating (Nie et al., 2021; Buswell et al., 2022).

1.8.4 Superstructure

Trade-offs exist when it comes to coating superstructures. Zinc/epoxy/polyurethane systems perform well and are common for new superstructure construction (LeBozec et al., 2018; Ghahari et al., 2018).

1.8.5 Underwater Hulls

While coatings on hulls must protect the underlying steel from corrosion, they must also keep it smooth (George et al., 2022). That is because any roughness on the hull creates favourable conditions for the thousands of species of plants and animals in oceans to become attached.

1.8.6 Exterior Hull (Boot-Topping)

The coating used to protect the hull usually is extended upward to the rail; antifouling coatings are not necessary on surfaces that won't come into contact with water (Tian et al., 2021).

REFERENCES

Abbas, M., & Shafiee, M. (2020). An overview of maintenance management strategies for corroded steel structures in extreme marine environments. *Marine Structures*, *71*, 102718.

Barboza, L. G. A., Cózar, A., Gimenez, B. C., Barros, T. L., Kershaw, P. J., & Guilhermino, L. (2019). Macroplastics pollution in the marine environment. In *World seas: An environmental evaluation* (pp. 305–328). Academic Press.

Bates, B. L., Zhang, L. Z., & Zhang, Y. (2015). Electrodeposition of Ni matrix composite coatings with embedded CrAlY particles. *Surface Engineering*, *31*(3), 202–208.

Bolli, E., Alessandra, F., Poato, F., Salius, K., & Mezzi, A. (2020). Cr segregation and impact fracture in a martensitic stainless steel. *Coatings*, *10*(9), 843.

Buswell, R. A., Bos, F. P., Silva, W. R. L. D., Hack, N., Kloft, H., Lowke, D., ... & Roussel, N. (2022). Digital fabrication with cement-based materials: Process classification and case studies. *Digital fabrication with cement-based materials: State-of-the-art report of the RILEM TC 276-DFC* (pp. 11–48).

Costa, J. M., & de Almeida Neto, A. F. (2020). Ultrasound-assisted electrodeposition and synthesis of alloys and composite materials: A review. *Ultrasonics Sonochemistry, 68*, 105193.

Cui, X., Zhu, G., Pan, Y., Shao, Q., Dong, M., Zhang, Y., & Guo, Z. (2018). Polydimethylsiloxane-titania nano composite coatings; fabrication and corrosion resistance. *Polymer, 138*, 203–210.

Fayomi, O. S. I., Popoola, A. P. I., & Olorunniwo, O. E. (2016). Structural and properties of Zn-Al_2O_3-SiC nano-composite coatings by direct electrolytic process. *The International Journal of Advanced Manufacturing Technology, 87*, 389–398.

George, J. M., Kimiaei, M., Elchalakani, M., & Efthymiou, M. (2022, September). Flexural response of underwater offshore structural members retrofitted with CFRP wraps and their performance after exposure to real marine conditions. In *Structures* (Vol. 43, pp. 559–573). Elsevier.

Ghahari, S. A., Ha, P., Chen, S., Alinizzi, M., Agbelie, B., & Labi, S. (2018). Inputs for bridge painting decision support: A synthesis. *Infrastructure Asset Management, 5*(2), 56–74.

Guedes Soares, C., & Garbatov, Y. (2011). Effect of environmental factors on steel plate corrosion under marine immersion condition. *Corrosion Engineering Science and Technology, 46*(4), 524–541.

Ibrahim, M. A., & Bakdash, R. S. (2020). Copper-rich Cu–Zn alloy coatings prepared by electrodeposition from glutamate complex electrolyte: Morphology, structure, microhardness and electrochemical studies. *Surfaces and Interfaces, 18*, 100404.

Jabbar, A., Yasin, G., Khan, W. Q., Anwar, M. Y., Korai, R. M., Nizam, M. N., & Muhyodin, G. (2017). Electrochemical deposition of nickel graphene composite coatings: Effect of deposition temperature on its surface morphology and corrosion resistance. *RSC Advances, 7*(49), 31100–31109.

Javidan, F., Heidarpour, A., Zhao, X. L., & Minkkinen, F. (2016). Application of high strength and ultra-high strength steel tubes in long hybrid compressive members; experimental and numerical investigation. *Thin Walled Structures, 102*, 273–285.

Jouyandeh, M., Rahmati, N., Movahedifar, E., Hadavand, B. S., Karami, Z., Ghaffari, M., ... & Saeb, M. R. (2019). Properties of nano-Fe3O4 incorporated epoxy coatings from Cure Index perspective. *Progress in Organic Coatings, 133*, 220–228.

Jumbo, E. E., Obiga, O., Obaseki, E., & Siongo, J. R. (2020). Predictive evaluation of corrosion signatures in materials selection and fabrication for deployment in marine environments. *Environments, 6*(4), 1–8. www.jmess.org.

Kirchgeorg, T., Weinberg, I., Hörnig, M., Baier, R., Schmid, M. J., & Brockmeyer, B. (2018). Emissions from corrosion protection systems of offshore wind farms: Evaluation of the potential impact on the marine environment. *Marine Pollution Bulletin, 136*, 257–268.

LeBozec, N., Thierry, D., & Pelissier, K. (2018). A new accelerated corrosion test for marine paint systems used for ship's topsides and superstructures. *Materials and Corrosion, 69*(4), 447–459.

Lehmhus, D., Busse, M., Herrmann, A., & Kayvantash, K. (Eds.). (2013). *Structural materials and processes in transportation*. John Wiley & Sons.

Li, J., Du, A., Fan, Y., Zhao, X., Ma, R., & Wu, J. (2019). Effect of shot-blasting pretreatment on microstructures of hot-dip galvanized coating. *Surface and Coatings Technology, 364*, 218–224.

Li, Y., Xu, T., Wang, S., Yang, J., Li, J., & Xu, D. (2019). Characterization of oxide scales formed on heating equipment in supercritical water gasification process for producing hydrogen. *International Journal of Hydrogen Energy*, *44*(56), 29508–29515.

Liu, Z., Zhao, D., Wang, P., Yan, M., Yang, C., & Chen, Z. (2022). Additive manufacturing of metals microstructure evolution and multistage control. *Journal of Materials Science & Technology*, *100*, 224–236.

Loto, C. A., Loto, R. T., & Popoola, A. P. I. (2011). An investigation of carburization resistance performance of ethylene furnace tube alloys. *International Journal of Physical Sciences (IJPS)*, *6*(19), 4602–4613.

Loto, R. T., & Loto, C. A. (2019). Evaluation of the localized corrosion resistance of 316 L austenitic and 430Ti ferritic stainless steel in aqueous chloride/sulphate media for application in petrochemical crude distillation units. *Materials Research Express*, *6*(8), 086516.

Mai, J., Yang, T., & Ma, J. (2022). Novel solar-driven ferrate (VI) activation system for micropollutant degradation: Elucidating the role of Fe (IV) and Fe (V). *Journal of Hazardous Materials*, *437*, 129428.

Mazare, A., Anghel, A., Surdu-Bob, C., Totea, G., Demetrescu, I., & Ionita, D. (2018). Silver doped diamond-like carbon antibacterial and corrosion resistance coatings on titanium. *Thin Solid Films*, *657*, 16–23.

Meng, X., Zeng, P., Lin, S., Wu, M., Yang, L., Bao, H., ... & Sun, W. (2023). Deep removal of fluoride from tungsten smelting wastewater by combined chemical coagulation-electrocoagulation treatment: From laboratory test to pilot test. *Journal of Cleaner Production*, *416*, 137914.

Mohanty, D., Barman, T. K., & Sahoo, P. (2019). Characterisation and corrosion study of electroless nickel-boron coating reinforced with alumina nanoparticles. *Materials Today: Proceedings*, *19*, 317–321.

Mohanty, S., & Gokuldoss Prashanth, K. (2023). Metallic coatings through additive manufacturing: A review. *Materials*, *16*(6), 2325.

Naveed, A., Rasheed, T., Raza, B., Chen, J., Yang, J., Yanna, N., & Wang, J. (2022). Addressing thermodynamic Instability of Zn anode: Classical and recent advancements. *Energy Storage Materials*, *44*, 206–230.

Nevshupa, R., Martinez, I., Ramor, S., Arredondo, A. (2018). The effect of environmental variables on early corrosion of high-strength low-alloy mooring steel immersed in sea water. *Marine Structures, 60*, 226–240.

Nie, W., Wang, D., Sun, Y., Xu, W., & Xiao, X. (2021). Integrated design of structure and material of epoxy asphalt mixture used in steel bridge deck pavement. *Buildings*, *12*(1), 9.

Nwuzor, I. C., Idumah, C. I., Nwanonenyi, S. C., & Ezeani, O. E. (2021). Emerging trends in self-polishing anti-fouling coatings for marine environment. *Safety in Extreme Environments*, *3*, 9–25.

Park, J. H., Ko, K. P., Hagio, T., Ichino, R., & Lee, M. H. (2022). Effect of Zn-Mg interlayer on the corrosion resistance of multilayer Zn-based coating fabricated by physical vapor deposition process. *Corrosion Science*, *202*, 110330.

Portela, D. G., de Morais Nepel, T. C., Costa, J. M., & de Almeida Neto, A. F. (2020). Two-stages electrodeposition for the synthesis of anticorrosive Ni–W–Co coating from a deactivated nickel bath. *Materials Science and Engineering: B*, *260*, 114611.

Rashid, K. H., & Khadom, A. A. (2022). Sulfosalicylic/oxalic acid anodizing process of 5854 aluminum-magnesium alloy: influence of sealing time and corrosion tendency. *Results in Chemistry*, *4*, 100289.

Sun, Y., Cheng, S., Mao, Z., Lin, Z., Ren, X., & Yu, Z. (2020). High electrochemical activity of a Ti/SnO2–Sb electrode electrodeposited using deep eutectic solvent. *Chemosphere*, *239*, 124715.

Tian, X., Sun, X., Liu, G., Xie, Y., Chen, Y., & Wang, H. (2021). Multi-objective optimization of the hull form for the semi-submersible medical platform. *Ocean Engineering, 230*, 109038.

Toklu, E., Gur, M., & Celtik, M. (2018). Investigation on effects of steel surface properties on galvanization behavior. *Journal of Engineering Research and Applied Science*, *7*(1), 861–868.

Ünal, E., & Karahan, I. H. (2018). Effects of ultrasonic agitation prior to deposition and additives in the bath on electrodeposited Ni-B/hBN composite coatings. *Journal of Alloys and Compounds*, *763*, 329–341.

Vieira, R. R., Duarte, I. D., Eleutério, H. L., de Oliveira, T. G., Bagatini, M. C., & Tavares, R. P. (2022). Trajectory of top-dross particles during the melting of zinc ingot in galvanizing pot. *Materials Research*, *25*, e20210406.

Walsh, F. C., Wang, S., & Zhou, N. (2020). The electrodeposition of composite coatings: Diversity, applications and challenges. *Current Opinion in Electrochemistry*, *20*, 8–19.

Wei, Y. K., Li, Y. J., Zhang, Y., Luo, X. T., & Li, C. J. (2018). Corrosion resistant nickel coating with strong adhesion on AZ31B magnesium alloy prepared by an in-situ shot-peening-assisted cold spray. *Corrosion Science*, *138*, 105–115.

Xia, L. (2013). *Preparation and formulation of powder coating for application of enamelled wire*. The University of Western Ontario (Canada).

Yang, H., Huang, Y., Song, B., & Kainer, K. U. (2019). Enhancing the creep resistance of AlN/Al nanoparticles reinforced Mg-2.85 Nd-0-92 Gd.0.41 Zn-0.29 Zn alloy by a high sheer dispersion technique. *Materials Science and Energy A*, *755*, 18–27.

Yu, Z., Hu, J., & Meng, H. (2020). A review of recent developments in coating systems for hot-dip galvanized steel. *Frontiers in Materials*, *7*, 74.

Zhang, H., Wang, W. X., & Chang, F. (2021). Microstructure manipulation and strengthening mechanisms of 40Cr steel via trace TiC nanoparticles. *Materials Science and Engineering A*, *822*, *141693.*

Zhang, X., Zhou, Q., Wang, K., & Peng, Y. (2019). Study on microstructure and tensile properties of high nitrogen Cr-Mn steel processed by CMT wire and arc additive manufacturing. *Materials and Design*, *166*, 107611.

Zubielewicz, M., Langer, E., Królikowska, A., Komorowski, L., Wanner, M., Krawczyk, K., ... & Hilt, M. (2021). Concepts of steel protection by coatings with a reduced content of zinc pigments. *Progress in Organic Coatings*, *161*, 106471.

2 Electrodeposition

Effects of Zinc Multifaceted Barrier Protection in Sustaining Engineering Structure

2.1 INTRODUCTION

Over the past decades, coating industries have undergone increased technology and structural modifications. These changes occur as a result of restrictions on the use of toxic and hazardous chemicals. However, further changes have occurred as competition to produce environmentally friendly coatings is on the increase without considering the application. The result of these changes led to the removal of previously well-established coating by the newer coatings. These formulations are based on new coating technology. Electrodeposition is referred to as a method of developing or modifying new or existing materials via coatings or thin films approach. The process of electrodeposition is associated with the reduction or deposition of the electroactive and this would be followed by a new species on the surface of the cathode (Akinfenwa et al., 2023a). The importance of the electrodeposition process is the fact that it makes the process a controllable one because of the application of the chemical process in different applications. Electrodeposition has been a viable method in coating applications (Kannan et al., 2022).

Electrodeposition finds applications in the formation of thin films used in thermoelectric products. Nanocables are produced via the electrodeposition process and this is one reason why electrodeposition is categorized into direct and pulse electrodeposition. The former usually forms a dendritic pattern causing a non-homogenous surface. However, the non-homogenous issues can be handled using the pulse deposition method, where the ions are diffused on the surface of the electrode and cause the formation of crystals on the alloyed material (Kumar & Singh, 2021). There are various types of electrodeposition processes, which include electroplating, electrophoretic, and underpotential positions. However, different failure modes evolve during the process, which includes the formation of the dendrite (in the case of Zn), hydrogen gas evolution wear, corrosion issues, and variation in morphology. For instance, in the plating of the Zn anode, a major problem occurs when the Zn anode cannot dissolve completely in the given electrolyte. This could cause undesirable problems in the process and lower the Zn metal stripping. Thus, the gradual dissipation of Zn on the surface of the anode takes place, resulting in the failures associated with batteries (Zheng et al., 2021). Electrodeposition is an

DOI: 10.1201/9781003562078-2

important technique in the production of key materials for manufacturing electronic chips and some storage systems requiring magnetic properties. This is due to the low cost and performance improvement the technique provides for the information system. More so, nano and biotechnology are equally gaining future in the use of electrodeposition technique. It has made it easy for unique nano and biomaterials to be developed for industrial applications. Major properties of electrodeposition that made it deployable in nano, bio, and micro applications include its ability to integrate functional materials that are 3D-printed, use of room temperature especially when using water-electrolyte, and electrodeposition can be done from small to large geometries (Lin et al., 2022).

Permalloy is an alloy of nickel and iron that can magnetize and demagnetize. It requires high magnetic permeability, especially when used in the development of Tesla components. Difficulties exist in the composition of the alloy which will give the precise geometries, and alloy composition is a major problem in determining the composition during the electrodeposition of microelectronics. Without the required coating thickness and subsequent mechanical and chemical properties such as material wear and corrosion, the functionality of such microelectronic material would be jeopardized. Thus, industries involved in the development of components for products like computers, modern cars, aircraft, and medical equipment must examine the failure behaviours in terms of wear, corrosion, or material oxidation (Lee et al., 2021). One of the major failure responses of the electrodeposition technique is linked directly to the substrate adhesion and cohesion strength of the coating. The adhesive failure response is usually referred to as delamination while the cohesive failure response is known as coating cracking. Both degradation responses occur at the interface of the coating. Thus, the limitations of adhesion result in the external cracking of the coating at the interface (Mróz et al., 2016). A study showed that electrodeposition is considered a fast and most economical technique in thin film deposition due to its multi-component alloy deposition with reduced cost and failure while in an application. It was reported that the electrodeposition of copper and tin using an alkaline electrolyte bath as well as zinc in an acidic electrolyte bath gave a reduced power consumption in a single-step preparation. Thus, low and high-valence materials can be developed via the electrodeposition technique.

The application of electrodeposition techniques has been successful in different areas such as the production of different electrode materials for welding applications. Metal oxides can then be developed using cathodic electrodeposition. For instance, the electrochemical reduction of NO_3 would result in hydroxide materials which transform into transition metals after the annealing process, e.g. NiO. However, the anodic electrodeposition involves the oxidation of transition metals from a soluble state to a precipitated oxyhydroxide, e.g. NiOOH (Jing et al., 2021; Pu et al., 2021). More so, the application of electrodeposition in the manufacturing industry cannot be overemphasized despite their shortcomings as a result of the variation in the properties of the alloying elements. For instance, the study by Li et al. (2021) established that most alkali metals are associated with irregular dendritic growth patterns while charging due to their reduced potentials in their electrode. To curb this problem, it is possible to introduce alkali salt during the electrodeposition process, thereby developing a product that will be free from the growths of dendrites, which would be feasible from their microstructural examination. Similarly, to make the energy industries prominent, a study established that the application of the electrodeposition process has helped in the development of a low-cost nanocomposite with a catalyst that can have oxygen evolution properties. This process could happen faster than other processes such as hydrothermal synthesis. Such product is usually associated with high-tech future applications as well as sustainability in the energy technology industries (Taherinia et al., 2022).

2.2 PROSPECTS OF ELECTRODEPOSITION TECHNIQUE

Electrodeposition, otherwise known as electroplating, involves the deposition of materials on a surface (conductor) using salt or ionic solutions in a controlled parameter. This technique has been applied extensively in several applications, such as in the development of thin film for semiconductors such as capacitors, resistors, and inductors (Akinfenwa et al., 2023b). Several parameters are optimized during this process, which include temperature, current density, and the composition of the electrolyte. It is a unique method where process control can be used to obtain a uniform microstructure as well as excellent structural properties (Das & Dhara, 2021).

For instance, C. Liu et al. (2022) applied a controllable electrodeposition technique using periodic nanostructures and a homogenous pattern of Au. Also, the template used showed that the material can be developed with evenness in its morphology and mechanical properties. Thus, the material developed was found to have applications in surface-enhanced Raman scattering. Fukami et al. (2022) reported that the production of lithium via electrodeposition usually creates the formation of dendrites just as it usually happens in the electrodeposition of zinc at the surface of the electrode. This has been one limitation of the application of lithium metals. Further to this, to solve this problem, it is possible to apply several additives to achieve a multifaced layer barrier coating. In the study, multi-component additives were used to simulate to mimic the actual electrodeposition process. Consequently, the results showed that the four additives added contributed different roles which enhance the applications of the metallic lithium. Wu et al. (2022) used an integrated approach for water treatment and purification by using multi-component electrode rotating equipment with electrodeposition under varying control parameters. The study employed a current density of about 53.1 A/m^2 and about 64 L/h, pH of 2, and a concentration value of Cd^{+2} was given as 800 mg/L. The result demonstrates that the removal efficiency of Cd^{+2} was about 99.4% in two hours (120 minutes) of electrodeposition. Thus, the study affirmed that under controlled electrodeposition, quality products, as well as quality, ensue. In the field of additive manufacturing, it has been established that the fabrication of 3D metallic and nanoscale structures has been challenging in additive manufacturing techniques. To this end, Wang, Xiong, et al. (2022) applied template electrodeposition in 3D printing to be able to fabricate the 3D metallic components. These structures were fabricated both at the microscale and nanoscale levels with excellent mechanical and structural properties.

Gong et al. (2022) utilized an ultrasonic electrodeposition technique in the deposition process of gold–tin alloys under a controlled parameter. The result showed that a gold–tin alloy with uniform thickness and density was developed, and it has been reported that the percentage of Sn based on the ultrasonic method lies between 30% and 60%. Also, the uniformity and density of the coating were traced to the application of the ultrasonic method, which improved the diffusivity of the ions. Alloys of tin and copper have wide applications in the production of bearings and essential components in electronics due to corrosion resistance properties and aesthetic visual characteristics. The electrodeposition technique finds greater application in this area because of the lower cost of production, flexibility, and quality surface finish. However, a major problem with such a process in this application is the incessant growth, which usually results in a coarse surface finish. Based on the limitations, Krajaisri et al. (2022) examined the influence of the current density and temperature on the electrodeposition of tin–copper alloys. The results showed that at higher temperatures, the coating properties were affected while all coating at pulse current densities and temperature exhibited less than 2%n Cu content. Thus, it is noteworthy to say that parameter controls enhance or help in the understanding of the electrodeposition process. Jiang et al. (2022) modified NiCoP using a jet-type electrodeposition technique in order to compare the result with the traditional electrodeposition method. The result showed that there was an improvement in the surface morphology as well as the surface adhesion ability of the NiCoP material. More so, the corrosion resistance of the NiCoP as well as the magnetic properties were observed to have improved. Thus, the magnetic properties of the developed NiCoP thin film material would do well in the development of magnetic materials.

Furthermore, Zhang et al. (2022) established that superalloy materials can be developed with strong corrosion resistance ability and prevented from embrittlement of hydrogen via the electrodeposition technique. However, while it is good for some industrial applications, it was reported that it has poor energy conservation as well as a reduction in the emission capacity. Thus, the addition of certain additives such as rare earth salt would help in the efficiency of the hydrogen inhibition during the electrodeposition process, thereby having great improvement in energy saving as well as reducing emissions during the electrodeposition, especially in the pickling applications. A study by Gao et al. (2012) provides hydrogen evolution issues associated with the electrodeposition process depending on the alloying elements involved during the process. For instance, gallium is in high demand in the field of information technology and it is found to be associated with several other elements like zinc aluminium and germanium, thus it cannot for a sole deposit of gallium without the aforementioned trace elements. However, during the electrodeposition process, there seems to be intense hydrogen evolution due to the low potential of gallium compared to the hydrogen evolution reaction. This could cause a concentration of Ga (3) in the electrolyte and eventual limitations in the mass transfer of Ga (3) ions leading to increased polarization (Xu et al., 2023). It was reported by Wang et al. (2021) that the electrodeposition technique helps in reducing the time required in developing some stable compounds or materials. For instance, the study used electrodeposition integrated with a sol–gel process to reduce the preparation time for the solutions. It was reported that the preparation time for the solution dropped from 12 hours to 30 minutes. This made it possible to deposit Cu on the Mo surface and ZnSnS sol–gel was mixed on the Cu surface to obtain CZTS films. Also, the developed material had improved mechanical and electrical properties as well as corrosion resistance properties. Thus, developing sustainable engineering materials could be developed via this approach.

Additionally, there are means of developing products having antibacterial properties; however, most products developed through coating techniques usually have poor durability owing to the adhesion issues between the coating and the substrates. To resolve this, Wang, Ye, et al. (2022) applied an electrodeposition technique in which a one-step pulse-reverse method was used to deposit Ag grains into the surface of stainless steel material (304 SS). The result demonstrates that a reliable product with high antibacterial reliability was obtained. Also, it was observed that there was an increase in the corrosion resistance as well as in surface abrasion of the developed material, thereby bringing innovation and sustainability to the manufacturing industry. Dai et al. (2022) carried out a one-step method for the preparation of MnO_2 electrodes for the electrodeposition of zinc ion batteries. The result showed that a porous structure reaction was formed, which paved the way for smooth electrochemical activity around the electrode. Thus, Coulomb efficiency, specific capacity, and current density were improved. It was further recommended that the one-step approach can be upgraded for commercialization. Thus, the prospect of electrodeposition, in general, is unlimited despite its several disadvantages, and many studies are yet to unravel the entire advantages of the technique. Also, a way of improving the manufacturing industries is through continual research for developmental purposes.

2.3 ZINC BARRIER COATING VIA ELECTRODEPOSITION ROUTE AND THEIR PROSPECTS APPLICATIONS

The zinc barrier coating electrodeposition technique has vast applications in the steel industry for protection, especially in structural reinforcement steels, aerospace, automobile, and electrical applications. However, at elevated temperatures, the corrosion protection devalues. Thus, alloying zinc via the electrodeposition process helps in increasing the tensile strength, corrosion resistance, and excellent mechanical properties compared to when zinc exists alone. The most utilized zinc alloy coatings are the zinc–nickel alloy owing to its corrosion performance, physical properties, and environmental friendliness. While different varieties of additives have been introduced into the electrodeposition of zinc–nickel alloy for excellent mechanical properties, El Boraei and Ibrahim (2022) added a small quantity of Sm_2O_3 nanoparticles into the sulphate bath used as the electrolyte during the deposition process. The results demonstrate that there was a paradigm shift in the polarization curve as well as in the deposition potential. This implies that the addition of the Sm_2O_3 nanoparticles gave a catalytic effect on the reduction of Zn^{2+} and Ni^{2+}. Furthermore, the addition of the nanoparticle to the zinc–nickel alloy deposition led to an increase in corrosion resistance and microhardness. More so, variations in the temperature and current densities were found to have led to variations in the morphology of the developed zinc–nickel coating. The results showed that there was variation in the polarization curve of the zinc–nickel alloy with the nanoparticle additive and without the additive.

The report indicated that the polarization curves and potentials moved to the right positively. Also, the addition of the nanoparticle is reflected in the increased value in the reduction current. Therefore, Sm_2O_3 nanoparticles have a great influence on the properties of the substrate surface. In addition, figures within the study also displayed the variation in their polarization curve as a result of the changes made in the pH of the solutions. It was observed that the polarization curves tend to move closer to a more noble potential region due to the increase in the pH of the bath solutions. Thus, the introduction of nanoparticles into the development of composites is now gaining prominence and zinc electrodeposition is playing a major role in the energy industry. Also, it is important to reiterate that the addition of nanoparticles under control parameters like current density and temperature has helped in developing complex shapes which would have been difficult to reinforce their mechanical while in solid form.

Shen et al. (2021) developed a coating having zinc, nickel, and cobalt in layers using an ultrasonic electrodeposition method. At a current density of about 1 A g^{-1}, the zinc doped with $Co(OH)_2$ as an inner layer and $Ni(OH)_2$ outer layer produced a specific capacitance of about 1774 F g^{-1}. The developed material was recommended for application in high-performance capacitors. Also, Huang et al. (2022) reported that electrodeposition of zinc via a rough interfacial approach causes some reaction at the electrolyte, which could result in an increase in internal reaction as well and limits its reversibility. This is a major problem facing the application of zinc metal commercialization, especially its anode. The study further proposed that integration of an interfacial approach to the layers gives a means of converting the zinc ion fluxes reaction by the formation of artificial layers. Thus, the electrodes are now made faster by the zinc redox action and increased current density would be achieved. It was also reported that there was stability in the thermodynamics of the metal anode and this caused avoidance of water-related side interfacial problems. Liu et al. (2021) deployed the potentiostatic electrodeposition approach to produce a uniform and dense zinc coating under controlled parameters like temperature and deposition potentials. The properties evaluated like the density, viscosity, and conductivity were observed to have improved against the as-received material. Also, the X-ray diffraction analysis (XRD) as well as scanning electron microscopy/energy dispersive spectroscopy (SEM/EDS) showed that the product from the cathode region confirmed that it is zinc coating. More so, the density and uniformity in the zinc coating were observed from the result of the microstructural evaluation.

According to Saidi et al. (2020), zinc metals find wide applications in the biological field; however, a major problem faced in the biological applications of zinc lies in the robust coating process. This is because it is difficult to control the parameters during the electrodeposition process. Based on this, the study leveraged the electrodeposition of zinc oxide on Ti6Al4V using a nitrate bath containing zinc. The experiment was controlled to obtain different microstructures having spherical, nano-like and worm-like patterns. The result equally demonstrates that the grain structure was modified from polygon to basal orientation due to the influence of temperature increase. Lower temperature produces zinc oxide via the initial growth of substrate nuclei. Thus, temperature and potentials played a major role in the electrodeposition of zinc oxide for efficient applications in the biological field.

2.4 FAILURE RESPONSE OF DEPOSITED ZINC ALLOYS

A major problem of orthopaedic equipment/devices is the presence of vanadium and aluminium and these elements affect the durability in a way. Electrodeposition of zinc oxide coating has been reported to have enhanced the corrosion resistance of Ti6Al4V. This was also complemented by the various morphological patterns of porous and worm-like types. Both morphologies were reported to have performed excellently well in their corrosion resistance ability (Saidi et al., 2020). Lau et al., (2022) reported that the highway bridges in some parts of the United States are failing since their major components are made of steel. Corrosion activities degrade the structural properties of the steel and eventually reduce its performance before its design life elapses. It was reported in the study that about 15% of the bridges in the United States are bad due to the corrosion actions of the steel material and this amounted to a cost of about $8.3 billion annually. However, while zinc-rich primer and other coating types were deployed in investigating their corrosion protection performance, the result showed that the electrochemical noise resistance was possible via the zinc and steel interface but exhibited good coating quality for the steel material.

To obtain a strong zinc multifaced layer barrier coating capable of giving good protective protection, there is a need for the choice of method of preparation as well as the parameter controls. For instance, Wang, Ye, et al. (2022) improved the mechanical and corrosion characteristics of zinc coating by integrating a hybrid material known as SiO_2-GO. It was reported that silicon oxide was introduced due to the need for the dispersion of the GO into the coating material. The results showed that there was a strong adhesion of about 0.7 MPa and the corrosion resistance increased from 0.4 wt% to about 24 wt%. This implies that the variation in the difference in corrosion resistance amounted to about 23.6%. Davoodi et al. (2020) improved the coating efficiency of zinc for construction and structural applications using rock types of minerals known as tectosilicate. The result showed that the developed material has strong corrosion resistance properties as well as strong adhesion properties. Haddadi et al. (2021) corroborated the above study by establishing that pigment addition to coating helps in improving the corrosion resistance and the overall mechanical properties of the material.

2.5 DESCRIPTION OF ENGINEERING FAILURES

Engineering components and structures are associated with failures due to the dynamic effect of the forces or loading. These loadings cause fatigue which may result in initial crack propagation on the surface of the component. However, over time, as the cyclic loading varies with stress/strain and time, fatigue sets in leading to larger cracks formation. This will eventually result in complete damage to the material. Failure could be in the form of chemical (corrosion), mechanical (wear), or electrical failure, as much as the failed component would either lead to stop or distortion of the machine operation. One of the importance of failure analysis is that it helps in the collection and analysis of failure data, which will help in the determination of cause and corrective action required for restoration of failure (Masriera et al., 2015). Different types of failure modes exist in materials, which include ductile, brittle, and fatigue fracture. Other types of material failure modes include corrosion, wear, and distortion. Failure modes typically refer to the behaviour of a machine or component that will result in its eventual failure. A critical example of failure mode is the corrosion of mild steel, which may result in degradation in strength that will eventually lead to failure. A study by Ezzeldin et al. (2022) established that an increase in the wet or dry cycle increases the corrosion rate and other properties such as strength, ductility, and in most cases hardness in a period. More so, the corrosion failure will result in further corrosion problems, like pitting and cavity formation.

Ductile fracture implies excessive deformation of material even before fracture. Examples of ductile fracture include the fracture of metals, soft steel, and rubber material. B. Li et al. (2022) adopted a plasticity model as a ductile failure technique to predict the failure of the hub of a turbocharger turbine. This was necessary to prevent or reduce the incessant ductile failure usually experienced by the aero engine fans or turbine. The result revealed that it was possible to determine the burst speed of the rotating turbine using the plasticity material model technique for ductile failure prediction. Ductile failure of material under the influence of varying loads has been reported to have been inconsistent in its failure behaviour. This is because, for ductile materials, failure is strongly linked to the direction or axis of the loading (Chouksey & Keralavarma, 2022). Chiu and Srivastava (2022) established that ductile failure in hardened alloys is usually intergranular in nature because there is no uniform precipitation. This is always obvious from their microstructures. Thus, it is expected that the material without uniform precipitation will experience plastic heterogeneity in its grain boundaries resulting in low yield strength. Brittle materials are characterized by little or no plastic deformation before their eventual failure. These materials include but are not limited to polymers, glasses, ceramics, and metals. For brittle materials, crack propagation is usually through a process called cleavage with a unique brittle surface. Structural steel is used as reinforcement in bridges and other structures or machines. Different cases of failure exist, either due to over-usage or used beyond its expected lifespan. Brittle failure as well as the determination of fracture toughness is key to ensuring environmental and structural safety. Thus, verification of material toughness and impact is necessary for the determination of the brittle failure criterion (Sieber & Stroetmann, 2017). A study by Beygi et al. (2021) reported that the formation of the intermetallic compound during friction welding constitutes a major source of brittle fracture at the joints of dissimilar metals. For instance, aluminium and steel are dissimilar metals and there is a tendency for strength reduction since intermetallic compounds would always exhibit fracture propagation. According to Oie et al. (2021), the increasing dimensions in the geometries of buildings and other related structures have increased the components' thickness, thus the occurrence of brittle fracture is inevitable.

Fatigue fracture refers to the deformation of material due to repeated excessive loadings. It involves the formation of microcracks and these cracks propagate into larger ones under repeated cyclic loading leading to the eventual failure of the material. Thus, the process is slow with crack initiation after which the cracks become stable and grow resulting in rapid fracture. The study of fatigue behaviour in materials has become very important in the automotive and aerospace industries. This is due to that majority of the components are composite in fabricated form and there is a need for bonds at different joints. Thus, it becomes imperative to study the fatigue behaviour of these components under dynamic loading. For instance, Moreira et al. (2022) deployed a technique for investigating the fatigue/fracture behaviour of a composite joint. The results showed that it was possible to determine the material failure trend and when to replace it to avoid unavailability during the application. Zhai et al. (2019) investigated the creep and deformation mechanism of T92/Super304H dissimilar weld joints under varying stress of between 85 and 165 MPa at 923 K. The microstructural observations showed that failure occurred at the heat-affected zone where there is fine grain. Further to this, there was a transformation of intergranular fracture at different regions to transgranular fracture at the edge region due to variation in creep pressure.

Corrosion failure refers to a mechanism of material degradation as a result of the reaction of metals and alloys in a certain environment. For instance, rusting of metal when exposed to air. This may transform into all regions of the material and eventually result in the formation of a pit at the surface. Different types of corrosion failure modes include pitting corrosion, galvanic corrosion, stress corrosion cracking (SCC), uniform corrosion, crevice corrosion, intergranular corrosion, etc. H. Li et al. (2022) investigated the corrosion damage of a hydraulic control pipeline under oil recovery conditions to determine the incessant root cause of failure to propose a method for its failure correction. The crack morphology resulting from the SEM/EDS under varying temperature environments of between 250°C and 350°C indicates that there was corrosion fracture at 260°C. Further to this, SCC ensued which was transgranular in shape and the corrosion failure was eventually associated with the anodic SCC. External corrosion of oil and gas pipelines could be in the form of hydrogen embrittlement (HE), hydrogen-induced cracking, SCC, and microbiologically induced corrosion. A major factor affecting the corrosion of oil and gas pipelines results from the inadequate application of coal tar and fusion-bonded coatings. Also, certain parameters of coatings such as pH are measures which if not maintained could result in the corrosion failure of the oil and gas pipelines. More so, a major problem in analysing corrosion failure, especially in oil and gas pipelines is how and when the external corrosion occurs on the pipelines. The concept and conditions causing the development of these corrosion failures remain fully unanswered in the available literature. However, the steel pipe for the oil and gas is known to be buried inside a corrosive environment like soil and there seems to be corrosion damage based on the soil stress condition (Wasim & Djukic, 2022). This occurs as a result of the response of a material to surface abrasion caused by wear. The change resulting from wear could be the thickness of the coating, shape, and change in the dimension of the material geometry. Various wear failure modes could be adhesive, corrosive, abrasive, and surface fatigue. Major factors causing wear failure include design, material selection, method of manufacturing, contact area, lubrication, properties of the material, environment, and most times materials being stressed beyond their operating limit.

Pujiyulianto et al. (2022) investigated the wear behaviour of an impeller ring using the common failure analysis for materials and these include chemical composition, microstructural study, and hardness. The result showed that there was adhesion at the surface of the failed ring as well as a fracture on the surface. Further analysis showed that there was a region on the ring that acted as a stress concentration point. This is the reason behind the initial stress formation and the eventual failure results due to fatigue stress. Also, the chemical composition showed that the material did not meet the technical material and this caused a major issue during the material applications. Y. Liu et al. (2022) investigated the wear failure of a coated plain bearing in the aerospace and automotive industries. The method of analysis involves the use of stage life and whole life while the microscopic and macroscopic changes in the torque were compared. The result showed that the signal of the torque indicates the service life of the bearing. However, the accumulation of failures on the coated bearing material could cause sudden failure of the bearings. Thus, it was observed that abrasive wear, adhesion, and plastic deformation resulted in the complete failure of the coated plain bearing.

REFERENCES

Akinfenwa, O. D., Fayomi, O. S. I., Atiba, J. O., & Anyaegbuna, B. E. (2023a). Chemical and microstructural investigation of starch modified zinc oxide paired as a composite superhydrophobic coating for mild steel protection. *Chemical Papers*, *77*(12), 7643–7651.

Akinfenwa, O. D., Fayomi, O. S. I., Atiba, J. O., & Anyaegbuna, B. E. (2023b). Development of starch-modified titanium oxide paired with zinc powder using the electrodeposition technique as a composite superhydrophobic coating for mild steel. *The International Journal of Advanced Manufacturing Technology*, *130*(5), 2847–2854.

Beygi, R., Carbas, R. J. C., Barbosa, A. Q., Marques, E. A. S., & da Silva, L. F. M. (2021). A comprehensive analysis of a pseudo-brittle fracture at the interface of intermetallic of η and steel in aluminum/steel joints made by FSW: Microstructure and fracture behavior. *Materials Science and Engineering: A*, *824*, 141812.

Chiu, E., & Srivastava, A. (2022). Intergranular ductile failure of materials with plastically heterogeneous grains. *Materialia*, *23*, 101439.

Chouksey, M., & Keralavarma, S. M. (2022). Ductile failure under non-proportional loading. *Journal of the Mechanics and Physics of Solids*, *164*, 104882.

Dai, Q., Li, L., Hu, B., Jia, Y., Tu, T., Hoang, T. K., ... & Song, L. (2022). One-step preparation of MnO_2 electrode for secondary aqueous zinc ion batteries by electrodeposition. *Materials Today Communications*, *31*, 103578.

Das, S., & Dhara, S. (Eds.). (2021). *Chemical solution synthesis for materials design and thin film device applications*. Elsevier.

Davoodi, M., Ghasemi, E., Ramezanzadeh, B., & Mahdavian, M. (2020). Designing a zinc-encapsulated feldspar as a unique rock-forming tectosilicate nanocontainer in the epoxy coating; improving the robust barrier and self-healing anti-corrosion properties. *Construction and Building Materials*, *243*, 118215.

El Boraei, N. F., & Ibrahim, M. A. M. (2022). Catalytic impact of Sm_2O_3 nanoparticles on the electrodeposition of zinc-nickel alloy. *Materials Chemistry and Physics*, *285*, 126138.

Ezzeldin, I., El Naggar, H., Newhook, J., & Jarjoura, G. (2022). Accelerated wet/dry corrosion test for buried corrugated mild steel. *Case Studies in Construction Materials*, *17*, e01152.

Fukami, K., Sakurai, A., Tsujimoto, T., Yamagami, M., Kitada, A., Morimoto, K., ... & Murase, K. (2022). Macroscopically uniform and flat lithium thin film formed by electrodeposition using multicomponent additives. *Electrochemistry Communications*, *136*, 107238.

Gao, Y., & Liu, J. (2012). Gallium-based thermal interface material with high compliance and wettability. *Applied Physics A*, *107*, 701–708.

Gong, L., Zhao, F., Wang, Z., Sui, Q., Xu, S., Liu, B., ... & Chen, Z. (2022). Preparation of cyanide-free gold-tin alloys by ultrasonic-assisted pulse electrodeposition. *Materials Letters*, *317*, 132029.

Haddadi, S. A., Ghaderi, S., Sadeghi, M., Gorji, B., Ahmadijokani, F., Mahdavian, A. R., & Arjmand, M. (2021). Enhanced active/barrier corrosion protective properties of epoxy coatings containing eco-friendly green inorganic/organic hybrid pigments based on zinc cations/Ferula Asafoetida leaves. *Journal of Molecular Liquids*, *323*, 114584.

Huang, Z., Zhang, W., Zhang, H., Fu, L., Song, B., Zhang, W., ... & Lu, K. (2022). Uniform zinc electrodeposition directed by interfacial cation reservoir for stable Zn–I2 battery. *Journal of Power Sources*, *523*, 231036.

Jiang, W., Li, H., Lao, Y., Li, X., Fang, M., & Chen, Y. (2022). Synthesis and characterization of amorphous NiCoP alloy films by magnetic assisted jet electrodeposition. *Journal of Alloys and Compounds*, *910*, 164848.

Jing, M., Wu, T., Zou, G., Hou, H., & Ji, X. (2021). Nanomaterials for electrochemical energy storage. In *Frontiers of nanoscience* (Vol. 18, pp. 421–484). Elsevier.

Kannan, P. K., Chaudhari, S., Dey, S. R., & Ramadan, M. (2022). Progress in development of CZTS for solar photovoltaics applications. *Journal: Encyclopedia of Smart Materials*, *1*, 681–698.

Krajaisri, P., Puranasiri, R., Chiyasak, P., & Rodchanarowan, A. (2022). Investigation of pulse current densities and temperatures on electrodeposition of tin-copper alloys. *Surface and Coatings Technology*, *435*, 128244.

Kumar, R., & Singh, R. (Eds.). (2021). *Thermoelectricity and advanced thermoelectric materials*. Woodhead Publishing.

Lau, K., Permeh, S., Lawler, J., Fadden, M., Jones, C., & Ortiz, C. (2022, March). Evaluation of Corrosion Prevention and Control Programs for Reinforced Concrete Bridges in Marine Environments. In *AMPP CORROSION 1(1)*, 1–5.

Lee, P. T., Chang, C. H., Lee, C. Y., Wu, Y. S., Yang, C. H., & Ho, C. E. (2021). High-speed electrodeposition for Cu pillar fabrication and Cu pillar adhesion to an Ajinomoto build-up film (ABF). *Materials & Design*, *206*, 109830.

Li, B., Cui, Y., Liu, S., Liu, Y., Wang, X., & Ding, Z. (2022). Experimental and numerical study based on ductile failure for the tri-hub burst of turbocharger turbine. *Engineering Failure Analysis*, *138*, 106295.

Li, H., Liu, W., Chen, L., Fan, P., Dong, B., Ma, Z., & Wang, T. (2022). Corrosion crack failure analysis of 316L hydraulic control pipeline in high temperature aerobic steam environment of heavy oil thermal recovery well. *Engineering Failure Analysis*, *138*, 106297.

Li, H., Murayama, M., & Ichitsubo, T. (2021). Dendrite-free alkali-metal electrodeposition from contact-ion-pair state induced by mixing alkaline earth cation. *Cell Reports Physical Science*, *3*(6), 1–5.

Lin, S. H., Lefeuvre, E., Tai, C. H., & Wang, H. Y. (2022). Fabrication of high-performance non-enzymatic sensor by direct electrodeposition of nanomaterials on porous screen-printed electrodes. *Journal of the Taiwan Institute of Chemical Engineers*, *137*, 104386.

Liu, A. M., Guo, M. X., Shi, Z. N., Liu, Y. B., Liu, F. G., Hu, X. W., ... & Wang, Z. W. (2021). Physicochemical properties of 1, 3-dimethyl-2-imidazolinone– $ZnCl_2$ solvated ionic liquid and its application in zinc electrodeposition. *Transactions of Nonferrous Metals Society of China*, *31*(3), 832–841.

Liu, C., Wu, J., Wang, S., & Fang, J. (2022). Directional controllable electrodeposition growth of homogeneous Au nano-rampart arrays and its reliable SERS applications. *Journal of Electroanalytical Chemistry*, *909*, 116120.

Liu, Y., Ma, G., Zhu, L., Wang, H., Han, C., Li, Z., ... & Huang, Y. (2022). Structure–performance evolution mechanism of the wear failure process of coated spherical plain bearings. *Engineering Failure Analysis, 135*, 106097.

Masriera, L., Fernández, M., Marani, J., Antonel, G., Acosta, C., & Halabí, J. (2015). Scientific method applied to failure analysis on engineering components. *Procedia Materials Science, 8*, 117–127.

Moreira, R. D. F., de Moura, M. F. S. F., Silva, F. G. A., Reina, J. P. A., & Rodrigues, T. M. S. (2022). A simple strategy to perform mixed-mode I+II fatigue/fracture characterisation of composite bonded joints. *International Journal of Fatigue, 158*, 106723.

Mróz, K. P., Kucharski, S., Doliński, K., Bigos, A., Mikułowski, G., Beltowska-Lehman, E., & Nolbrzak, P. (2016). Failure modes of coatings on steel substrate. *Bulletin of the Polish Academy of Sciences. Technical Sciences, 64*(1), 249–256.

Oie, N., Kawabata, T., Kishiki, S., & Nakagomi, T. (2021). Specimen size effect on the brittle fracture properties of steel for buildings subjected to large earthquakes. *Procedia Structural Integrity, 33*, 586–597.

Pu, J., Zhang, K., Wang, Z., Li, C., Zhu, K., Yao, Y., & Hong, G. (2021). Synthesis and modification of boron nitride nanomaterials for electrochemical energy storage: from theory to application. *Advanced Functional Materials, 31*(48), 2106315.

Pujiyulianto, E., Muhyi, A., Paundra, F., Perdana, F., Yudistira, H. T., & Syaukani, M. (2022). Failure analysis of a wear ring impeller. *Engineering Failure Analysis, 138*, 106415.

Saidi, R., Ashrafizadeh, F., Raeissi, K., & Kharaziha, M. (2020). Electrochemical aspects of zinc oxide electrodeposition on Ti6Al4V alloy. *Surface and Coatings Technology, 402*, 126297.

Shen, C., Guan, X., Tang, Y., Zhao, X., & Zuo, Y. (2021). A zinc-cobalt–nickel heterostructure synthesized by ultrasonic pulse electrodeposition as a cathode for high performance supercapacitors. *Journal of Electroanalytical Chemistry, 902*, 115793.

Sieber, L., & Stroetmann, R. (2017). The brittle fracture behaviour of old mild steels. *Procedia Structural Integrity, 5*, 1019–1026.

Taherinia, D., Moazzeni, M., & Moravej, S. (2022). Comparison of hydrothermal and electrodeposition methods for the synthesis of $CoSe_2/CeO_2$ nanocomposites as electrocatalysts toward oxygen evolution reaction. *International Journal of Hydrogen Energy, 47*(40), 17650–17661.

Wang, J., Wang, Y., Li, H., Zhao, A., Li, B., Bi, J., & Li, W. (2021). Influences of alkali incorporation and electrodeposited metal layer on formation of MoS_2 in CZTS thin films. *Materials Science in Semiconductor Processing, 134*, 105943.

Wang, X., Ye, X., Zhang, L., Shao, Y., Zhou, X., Lu, M., ... & Bai, J. (2022). Corrosion and antimicrobial behavior of stainless steel prepared by one-step electrodeposition of silver at the grain boundaries. *Surface and Coatings Technology, 439*, 128428.

Wang, Y., Wu, M., Lu, P., Zhou, W., Shi, X., Yang, K., & Miao, X. (2022). Mechanical and corrosion resistance of cold sprayed zinc (CSZ) nano composite coating enhanced by SiO_2-GO hybrid material. *Colloids and Surfaces A: Physicochemical and Engineering Aspects, 632*, 127824.

Wang, Y., Xiong, X., Ju, B. F., & Chen, Y. L. (2022). Voxelated meniscus-confined electrodeposition of 3D metallic microstructures. *International Journal of Machine Tools and Manufacture, 174*, 103850.

Wasim, M., & Djukic, M. B. (2022). External corrosion of oil and gas pipelines: a review of failure mechanisms and predictive preventions. *Journal of Natural Gas Science and Engineering, 100,* 104467.

Wu, C., Gao, J., Liu, Y., Jiao, W., Su, G., Zheng, R., & Zhong, H. (2022). High-gravity intensified electrodeposition for efficient removal of Cd^{2+} from heavy metal wastewater. *Separation and Purification Technology*, *289*, 120809.

Xu, K., Mu, C., Zhang, C., Deng, S., Lin, S., Zheng, L., ... & Zhang, Q. (2023). Antioxidative and antibacterial gallium (III)-phenolic coating for enhanced osseointegration of titanium implants via pro-osteogenesis and inhibiting osteoclastogenesis. *Biomaterials*, *301*, 122268.

Zhai, X. W., Du, J. F., Li, L. P., Zhou, C., & Zhang, Z. (2019). Creep behavior and damage evolution of T92/Super304H dissimilar weld joints. *Journal of Iron and Steel Research International*, *26*(7), 751–760.

Zhang, P., Xu, Z., Zhang, B., Lei, B., Feng, Z., Meng, G., ... & Wang, F. (2022). Enhanced inhibition on hydrogen permeation during electrodeposition process by rare earth (RE= Ce) salt additive. *International Journal of Hydrogen Energy*, *47*(29), 13803–13814.

Zheng, X., Ahmad, T., & Chen, W. (2021). Challenges and strategies on Zn electrodeposition for stable Zn-ion batteries. *Energy Storage Materials*, *39*, 365–394.

3 Diesel Engine Bearings in Harsh Environments

The Strategic Role of Coating

3.1 INTRODUCTION

In material science, a coating is generally referred to as the formation of a layer of material over another for protection against degradation, and for decoration (Harsimran et al., 2021). Matcrials on which coatings are applied are known as substrates. A lot of substrate materials such as mild steel, stainless steel, aluminium, and wood have been coated for protection against corrosion and the enhancement of mechanical, microstructural, electrical properties, and other properties required in the material (Szubert et al., 2019; He et al., 2019). However, while some coatings have been found durable on these substrates, others have reportedly failed due to the wrong choice of coating and substrate materials, poor surface preparation, wrong choice of process parameters, an inappropriate coating method, and the application of the coating in the wrong environment different from which it was designed for (Miguel et al., 2019; Mehboob et al., 2020). For the coating of mild steel, some of the techniques used are laser alloying and cladding, hot dip (galvanizing), electroless plating, painting, electrodeposition, etc. (John et al., 2019).

From the literature, most of these coating techniques have been widely explored; however, electrodeposition is considered to be one of the most advantageous due to simplicity of operation, relatively low cost of production, high production rate, production of good quality thin films and good reproducibility, and easy control of pollution (Ghaziof & Gao, 2015; Blejan & Muresan, 2013). Literature has also described electrodeposition as a process that uses electric current to reduce dissolved metal cations to form a coherent metal coating on an electrode (Walsh & Low, 2016). With this technique of coating, it is envisaged that the challenges with bearings made of mild steel will be minimized or eliminated. Bearings degrade after a short period of usage due to exposure to an external contaminant, especially in an environment that is not eco-friendly (Wu et al., 2021). The degradation can be in the form of corrosion, which can further result in the loss of mechanical, microstructural, and other relevant properties desired in the bearing. The use of bearings on parts of engines and machinery helps in the reduction of friction between two sliding metallic surfaces. Most bearings are mainly used for supporting machinery shafts in rotation.

However, the normal force exacted by rubbing two bodies that are loaded does generate heat energy by the surface friction and wear, this heat energy is

DOI: 10.1201/9781003562078-3

generated, and some of the heat energy is absorbed by the material to create internal stress, which is debilitating to the bearing material, others are lost to the environment and makes the environment harsher (Shirazi et al., 2020). But with coatings on the bearing material, a substantial part of the heat energy does not come in contact with the bearing. The coating absorbs the heat and dissipates a reasonable portion part of it to the surroundings. Harsh environmental conditions and continuous contact with contaminating bodies could also aggravate material deterioration. Coating and designing of bearing with the right materials will improve hardness and minimize the wear rate and friction on the two sliding surfaces, which could result in the failure of the machinery. The most significant aim of the design of bearing is for the longevity of the bearing life in the machine, reduction of energy losses due to friction, reduction of bearing wear, and minimize the expensive/running cost of the engine or machinery due to maintenance, downtime of machinery as a result of constant failure of the main bearing. The unexpected bearing failure in manufacturing plants will cause a huge loss in the production process of the industry. In marine, automotive, and aviation industries, safety measures are often ensured to prevent unexpected failures in the main bearing, at all costs (Scheu et al., 2019).

Therefore, the design and selection process for the type of materials for coating the main bearing and bearing material is of utmost importance for the reliability of machinery operation. For bearing design, it is an important aspect of machine design that must be taken into consideration in the design process (Li et al., 2019). In machinery maintenance, most of the maintenance work that is done on the engines is on bearing lubrication. Failure to carry out proper lubrication might result in bearing failure and its constant replacement (Kareem, 2017). Thus, the selection of bearing substrate material type and form of coating will help to reduce the risks of bearing failure during the operation of the engine. The coating enhances the bearing fatigue cycle and bearing lifespan, since most coatings exhibit self-lubricating properties, which could aid the taking away of heat from the bearing's surfaces, therefore improving the lubricity between the bearing materials and the contact materials (Bashandeh et al., 2019).

Appropriate selection of coating materials for bearing application in a marine environment, aircraft, and automobiles is paramount for the safety of lives and engines, maintenance reduction cost, and the longevity of bearing. The type of load and loading cycle on bearings are also important factors to consider in the design and the selection of bearings. The operation of the bearing will result in the generation of heat energy. Hence, it is essential to select coating materials that can withstand and absorb the heat reduction generated to avoid material distortion or dislocation and sometimes high-temperature corrosion that could result from chemical attacks from gases and molten metals, usually at high temperatures (Pedeferri et al., 2018). Typically, pressure and temperature increase directly leads to an increment in corrosion rate because electrochemical reactions take place faster usually at high temperatures. An increase in temperature increases the energy of reactions, which surges the corrosion rate. To reduce this effect on coatings, biodegradable materials such as pulverized or nano-sized cow bones and other natural waste products containing fibres can be used as reinforcement in inorganic and metal coatings. These are introduced into the coating bath during the electrodeposition process in varying proportions and are co-deposited along with the other constituents of the bath. These materials also help to increase the load-carrying capacity of coatings (Wan et al., 2018).

Moreover, in the preparation of the coating bath, besides ensuring the correct choice of coating material, the right choice of process parameters such as coating temperature, time, concentration of constituents, pH, stirring rate, current density, and deposition voltage should be ensured. These parameters influence the performance characteristics of coatings (Kumar et al., 2016). For instance, coating at an appropriate temperature aids the dissolution of constituents of the bath. More so, a suitable choice of stirring rate could also partly help in the particle dissolution and largely assist in the electrophoresis and mass movement of particles to the substrate. Stirring of the bath constituents also prevents the agglomeration of particles (Unal & Karahan, 2018). Also, coating thickness can sometimes be proportional to the coating time. These process parameters often influence the mechanical, corrosion, microstructural, electrical, and other properties of coatings.

This section has briefly introduced coating and the effects it could have on bearing materials in marine and other degradation-prone industries. Coatings are generally employed to cover material surface against degradations and to also deposit some desired properties. These properties could be mechanical, microstructural, electrical properties, etc. Coating fails due to poor surface preparation, the wrong choice of coating and substrate materials, the wrong choice of process parameters, the application of the coating in a wrong environment different from what it was designed for, and the inappropriate coating method. The function ability of coatings could also be enhanced through the introduction of biodegradable materials such as pulverized or nano-sized cow bones and other natural waste products containing fibres that can be used as reinforcement in inorganic and metal coatings. These materials help to increase the load-carrying capacity of the coating. Coating largely improves lubricity and helps to absorb heat from the materials on which they are coated. The forthcoming sections will centre on bearings, exploring their various types and potential preventive measures against bearing failures.

3.2 BEARINGS

Machinery that rotates is used for the conversion of useful work for the production of power, driving of shafts for ship propulsion, and other uses. Journal bearings have been in use for driving rotating machinery and equipment. It consists of an inbuilt axle having little clearance that separates the journal from the bearing. The use of lubricating oil to fill this clearance helps reduce the effect of wear between the two surfaces either by operating in a hydrodynamic or a mixed regime (Gu et al., 2016). The main engine crankshaft and connecting rod are supported by the main bearing which transmits the combustion load from the piston connecting rod and crankshaft to rotational movement without metal-to-metal contact. Various forces are acting on the bearing during its operation. They are as follows:

- Gas pressure, which is created from the liner.
- Dynamic force, which is a result of different main engine rotational motions.

- Centrifugal forces, which are also a result of different rotational loads from the main engine.
- Frictional forces between the engine crankshaft and bearing as a result of the vibration of the main engine.

The main bearing can support and withstand different forces acting on it at a high speed of rotation without any issues. However, the choice of materials for its design is essential to achieve these objectives and also support the journal of the crankshaft even at any minor irregularities of the surface of the journal (Witek et al., 2017; Morris et al., 2018; Wu et al., 2021).

3.3 TYPES OF MAIN BEARING

The bearings used for the main engine for both power generation and propulsion are of three types, which are lead bronze bearing, bi-metal bearing, and tri-metal bearing.

3.3.1 Lead Bronze Bearing

The lead bronze bearing used for power generation and propulsion is made up of the following layers:

- Flash layer: This layer is found on the uppermost part of the bearing with a thickness of about 0.035 mm and the material used in its design is tin and lead. This layer is mainly used for corrosion and dust protection of the bearing surface, which flashes off when it's been in operation.
- Nickel barrier: This barrier is the second part of the bearing layer surface and the material used in its design is nickel. This layer has a thickness of 0.02 mm and this layer helps in the prevention of tin diffusion into the bearing metal and is also used in the prevention of corrosion of the bearing.
- Lead bronze layer: The lead bronze layer is the third part of the main engine bearing made up of lead bronze. The lead bronze layer helps in preventing bearing seizure due to its anti-seizure property used in the design of this layer. The lead bronze layer is the major layer and material used in this type of bearing.
- Steel back: This layer is the last layer of the bearing and it is in the back part of the bearing. The steel back surface gives the other layers of the bearing support and the needed shape and bonds together all the other layers (Changyun et al., 2013; Sharma et al., 2019).

3.3.2 Bi-metal Bearing

The bi-metal bearing is another type of bearing used in the main engine. It is made up of layers which are as follows:

- Aluminium tin: The aluminium tin of the bi-metal bearing is the first layer of the bearing having a thickness of 0.5–1.3 mm. The aluminium tin is the major layer of the bi-metal bearing.

- Bonding layer: The bi-metal bearing bonding layer is the second layer of the bi-metal bearing and it is made up of aluminium of about 0.1mm of thickness. This layer provides good bonding between the shell layer and the top layer of the bi-metal bearing.
- Steel back: The bimetal steel back gives the bimetal bearing the needed shape and support to the bearing which enables the bearing to withstand the various forces acting on it (Oksanen et al., 2017; Behrens et al., 2018; Zorn et al., 2020).

3.3.3 Tri-Metal Bearing

The tri-metal bearing is the third type of bearing used as the main engine bearing and it is made up of three layers, including the flash layer.

- Flash layer: This flash layer is the uppermost part of the tri-metal bearing made up of tin and lead with a thickness of one micron. This layer is used for corrosion and dust protection of the bearing surface. The flash layer flashes off as soon as the bearing is in operation.
- Over layer: The over layer is the second layer of the tri-metal bearing having a thickness of 20 microns and the material used in its design is white metal. This layer is the major component of the tri-metal bearing.
- Inter layer: The inter layer is the third layer of the tri-metal bearing having a thickness of 5 microns. This layer is used for corrosion prevention since its material is anti-corrosive.
- Lining: The lining is the layer of the tri-metal bearing between the inter layer and the steel back layer having a thickness of 1 mm and the materials used in its design are lead and bronze.
- Steel back: The steel back layer is the last layer of the tri-metal bearing mainly used for bearing support and shape (Kopeliovich, 2011; Chikalthankar et al., 2014; Summer et al., 2019).

3.4 PROPERTIES OF MAIN BEARING MATERIALS FOR MARINE DIESEL ENGINES

In the selection of the main bearing of engines, the features of bearing materials are as follows:

- Anti-corrosion properties prevent or reduce the rate of corrosion of the main bearing and other bearing parts.
- The bearing material should have resistance due to the friction of the surface.
- It should have a high capacity for load-bearing due to the dynamic loading of the main bearing when it is in operation.
- The oil film of the lubricating oil should be allowed by the bearing full proper journal rotation.
- The lubricating oil and the bearing material should not react with each other.

- The bearing material should have good embed-ability properties to prevent the damage to bearing surface due to little particles embedded in the bearing surface during the running of the engine.
- It should have good compressive and tensile strength.
- The bearing material should be good thermal resistance property due to the hotness of the bearing when it is in operation (Nagata et al., 2012; Gebretsadik et al., 2015; Brockett et al., 2017).

3.5 POSSIBLE BEARING DEFECTS

After some periods of application of bearings, the following defects are sometimes observed:

- Corrosion: When the surface of the main bearing is corroding, it causes the bearing surface to be dis-coloured and rough. This is a result of using acidic lubricating oil in the lubrication of the bearing. Corrosion of the bearing surface will lead to bearing failure with time.
- Abrasion: The surface of the bearing will be damaged with fine scratches when the engine is run with heavy fuel oil and the lubricating oil will be contaminated with particles. When the lubricating oil is not treated and filtered properly, these particles will be deposited on the surface of the hearing which will damage the bearing surface and subsequently affect the performance of the engine.
- Erosion: Stripping the over layer of the bearing material is most common in medium-speed engines. This is a result of insufficient lubricating oil supply pressure or unusual movement of the journal which causes the erosion of the bearing.
- Fatigue: Damage to the bearing lining is a result of too high engine load over the bearing capacity. This will exact fatigue on the bearing surface and subsequent failure of the bearing due to none functionality.
- Wiping: When the bearing is in operation, its temperature increases, which takes off the bearing over layer. The process of removal of over layer helps in re-aligning the bearing to the bearing journals. But, when the wiping of the bearing metal is much, it will result in an increase in the bearing clearance, which will affect the bearing performance (Fonte et al., 2015; Farfan-Cabrera & Gallardo-Hernandez, 2017; Lee et al., 2019; Kvryan et al., 2020).

3.6 POTENTIAL FAILURE MODES OF BEARINGS

The best way of examining the failure of a journal bearing is through the sleeve surface to determine how the bearing material failed. Each mode of failure does have several root causes that lead to the failure of the surface of the bearing. The main engine bearing modes are discussed as follows:

- Fatigue: This mode of failure is characterized by cracks on the surface of the bearing and areas where the overlay flakes off. This mode of failure is usually caused by the forces acting on the bearing exceeding the fatigue strength of the design material. As the flake off of the overlay continues, the load acting on the bearing is concentrated on the bearing extruding surfaces, thereby causing the bearing surface wear to be accelerated.

- Wiping: This mode of failure is a result of the increasing temperature of the surfaces of the bearing when the bearing is in operation. This increasing temperature can melt the overlay babbitt material. This melted material will then be deposited on a cooler area on the bearing surface.
- Scoring: This is a result of deep radial scratches on the surface of the bearing due to foreign particle contamination being embedded on the overlay surface of the bearing. These foreign particles can be from the supply of lubricating oil to the bearing and wear particles of the bearing materials. It creates a high spot on the surface of the bearing which can cause the metal-to-metal wearing of the surface of the journal bearing.
- Corrosion: The surface of a bearing can be affected by a chemical attack and the processes of oxidation do result in corrosion on the surface of the bearing. When the surface of the bearing is being corroded, it will result in failure of the bearing due to fatigue, wear, and scoring of the surface of the bearing. Bearing corrosion is usually caused by lubrication oil contamination with either coolant or water and also inappropriate changeover of lubricating oil (Pan & He, 2015; Repka et al., 2017; Kvryan et al., 2020).

3.7 BEARING SURFACE COATING

The use of coating on the surface of the main bearing is to eliminate bearing surface wearing as a way of increasing the lifespan of the bearing. Bearings are coated to improve their performance when used in the running of diesel engines and also reduce the wear rate of the bearing when using suitable coating materials on the substance of the cast iron. Coatings on the bearing surface are sometimes deposited on the surface of the substrate in a very thin layer. When the deposit on the substrate is very thin about a few microns, it has the advantage of tolerance that is unchanged during machining.

3.7.1 Electroplating

The temperature of a bearing for a rolling mill is from about 100°C to 200°C from the start of an operation to the stop of the operation. In other to improve the bearing lifespan and bearing fatigue, the surface can be coated with ceramics added to the coating material (Ni/SiC and Ni/Al_2O_3) to improve the surface wear rate. The substrate is the cathode while the coating metal/material is the anode. Each of the materials to be used for the process must have a contact surface that is stable. For better performance on the coating surface, the use of shielding plates and good scattering electrolytes and the thickness of the layer for bearings are less than 5 microns (Goral & Skrzypek, 2018; Ma et al., 2020; Zhang et al., 2021).

3.7.2 Magnetron Sputtering

The process is environmentally friendly. It uses dense film and a thin metallographic structure in a vacuum. This process of bearing preparation gives the bearing performance superiority over the electroplated bearing surface coating. The coefficient of friction was about 0.1–0.25 when CuSnSb film was prepared on a steel substrate by magnetron sputtering experiment with the bearing running at about 4000 rpm. When Cu film was sputtered with Ti3SiC2 material, the bearing showed great improvement in its physical and mechanical properties (Yang et al., 2018).

3.7.3 Liquid Dope Coating

The liquid dope spraying coating method is another coating method for bearing surface coating; the process is environmentally friendly with high efficiency in terms of delivery. The use of MoS_2 and Sb_2O_3 with epoxy resin in coating the bearing surface improves the wear rate of the bearing surface and life span. This combination can withstand bearing operation temperatures above 200°C (Jia et al., 2015; Borgaonkar et al., 2021).

3.7.4 Power Sintering Metallic Coating

This process of bearing coating is usually done with a mixture of metallic powder with other functional powder, which will be placed in a high-temperature furnace for the composition. This method increases the bearing performance and reduces the wear rate of the bearing when it is in operation (Jarząbek et al., 2017; Chen et al., 2018).

3.7.5 Thermal Spraying Coating

This method uses supersonic flame spraying technology in the spraying of the molten alloy powder at about 500 m/s to the bearing surface. The powder was first heated to melt the powder at a high temperature before it could be used for coating the bearing surface. The use of mixed powdered Cr_3C_2, NiCr, and CaF_2/BaF_2 at various percentages (54%, 34%, and 12%, respectively) for the supersonic flame spraying will give a friction coefficient of 0.75 and 0.37 of the bearing when it is running at temperatures of 20°C and 500°C, respectively due to this mixture of coating on the bearing surface (Huang et al., 2009; Geng et al., 2016; Shakhova et al., 2017).

3.7.6 Supersonic Cold Spray Coating

This technology is a recent technology developed from supersonic hot spray coating technology, which uses low temperature at high speed in the coating process. The use of $CuSn_6$, $CuSn_8$, $CuSn_{45}$, $AlSn_5$, and $AlSn_{20}$ materials in the cold spray coating process will give a coefficient of friction higher than 0.5 with excellent elastoplasticity when the bearing is in operation (Hofsajer et al., 2013; Cao et al., 2019).

3.7.7 Polytetra Fluoroethylene (PTFE) Composition Coating

The PTFE coating process cannot be bound together easily with the substrate surface. It can use either the suitor bronze composite, PTFE fabric or the use of PTFE composite in the bounding process. The sinter bronze composition is made up of steel backing, from copper of about 0.3 mm thick. The overlay of this type of bearing is about 10–30 mm thick, which is formed from the PTFE and MoS2 filled to the pores of the suitor bronze layer lattice. That of the PTFE fabric is woven to form a double-face fabric. The basic linear material is about 0.6 mm thick, and is made from the impregnation of synthetic resin with fibres and glass fibres; the product is then laminated to form it. That of the PTFE composite is formed from the addition of PTFE additives and polyamide-reinforced glass fibre. The load and temperature of this technique are less than that of the sinter bronze composite and the PTFE fabric. This bearing coating process is mainly used for the prevention of sticky slip and fretting on bearing racy and its maintenance-free sliding contact area (Demirci & Duzcukoglu, 2014; Rudnev et al., 2016; Dhanumalayan et al., 2018).

3.7.8 Physical Vapour Deposition

The process involves clearing the surfaces by argon ion bombardment. The sputtering source is supplied by a high negative potential gas discharge which evaporates the material. The reactive gas moves this evaporated coating of the sputtered particles and deposits it on the surface of the substrate. This technology normally deposits very thin layers of metal and carbon on the substrate surface (Baptista et al., 2018; Mehran et al., 2018).

3.7.9 Ion Plating

The ion plating coating flux is provided by thermal evaporating in which the atom has been evaporated to the plasma region for ionization. The deposition of the film on the substrate is accelerated with an electric field before it has been deposited (Niu et al., 2015; Park et al., 2015).

3.7.10 Activated Reactive Evaporation

The process is usually used for the nitrides, carbides, and metal oxide deposition on a substrate. The reaction between the reactive gas atom and that of evaporated atoms takes place in a plasma region. The process is achieved with the injection of the reaction gas into the plasma region (Santecchia et al., 2015; Mehran et al., 2018).

3.7.11 Ionized Cluster Beam Deposition

The evaporation process of this technology is through a nozzle from a closed source. A cluster of atoms is formed when atoms expansion through the nozzle is cooled, which is about 100–1000 atoms per cluster. The evaporated atom is accelerated to the substrate after its ionization through the plasma region and the energy used for its acceleration is about 0.2 eV and above. The lattice damage on the surface of the substrate is reduced as a result of the low energy used in the process (Qu et al., 2015; Piazzoni & Milani, 2019).

3.7.12 Thermally Activated Chemical Vapour Deposition (CVD)

The vaporization of a volatile compound decomposed thermally or reacts with other gases, liquids, or vapour will produce non-volatile products deposited on the substrate automatically. The reaction uses a few Torr (Torr equals 1 mmHg) above atmospheric pressure and temperature of about 1000°C for the process (Carminati et al., 2018; Yang et al., 2018).

3.7.13 Plasma-Assisted CVD

This technology uses ions and electrons that travel through the reactor containing neutrals in which the electric power supplied to the reactor creates an electric field in the plasma region and takes a chemical reaction between the neutrals, ions, and electrons. The working temperature, pressure, and energy developed in the process are 23,000–92,800 K, 10–100 Pa, and 2–8 eV, respectively. Thermal equilibrium between the electrons and neutral gas is not achieved due to the temperature difference that exists in the molecules (Vetter et al., 2005; Moreno-Couranjou et al., 2018).

REFERENCES

Baptista, A., Silva, F. J. G., Porteiro, J., Míguez, J. L., Pinto, G., & Fernandes, L. (2018). On the physical vapour deposition (PVD): Evolution of magnetron sputtering processes for industrial applications. *Procedia Manufacturing*, *17*, 746–757.

Bashandeh, K., Lan, P., & Polycarpou, A. A. (2019). Tribological performance improvement of polyamide against steel using polymer coating. *Tribology Transactions*, *62*(6), 1051–1062.

Behrens, B. A., Goldstein, R., & Chugreeva, A. (2018). Thermomechanical processing for creating bi-metal bearing bushings. *Thermal Processing in Motion*, 15–21.

Blejan, D. M. L. M., & Muresan, L. M. (2013). Corrosion behavior of Zn-Ni-Al_2O_3 nanocomposite coatings obtained by electrodeposition from alkaline electrolytes. *Materials and Corrosion*, *64*(5), 433–438.

Borgaonkar, A. V., Syed, I., & Sonawane, S. H. (2021). Effects of lubrication on tribological properties coating material of composite MoS_2-TiO_2. *Tribological Applications of Composite Materials*, *1*, 267.

Brockett, C. L., Carbone, S., Fisher, J., & Jennings, L. M. (2017). PEEK and CFR-PEEK as alternative bearing materials to UHMWPE in a fixed bearing total knee replacement: An experimental wear study. *Wear*, *374*, 86–91.

Cao, C., Li, W., Zhang, Z., Yang, X., & Xu, Y. (2022). Cold spray additive manufacturing of Ti6Al4V: special nozzle design using numerical simulation and experimental validation. *Coatings*, *12*(2), 210.

Carminati, P., Buffeteau, T., Daugey, N., Chollon, G., Rebillat, F., & Jacques, S. (2018). Low pressure chemical vapour deposition of BN: Relationship between gas phase chemistry and coating microstructure. *Thin Solid Films*, *664*, 106–114.

Changyun, L., Ying, W., Jinzhi, Z., & Guofa, M. (2013). Manufacturing of lead bronze bearing by centrifugal casting and its microstructure and properties. *Special Casting & Nonferrous Alloys*, *7*, 653–655.

Chen, A. N., Wu, J. M., Liu, K., Chen, J. Y., Xiao, H., Chen, P., ... & Shi, Y. S. (2018). High-performance ceramic parts with complex shape prepared by selective laser sintering: A review. *Advances in Applied Ceramics*, *117*(2), 100–117.

Chikalthankar, S. B., Nandedkar, V. M., & Jawale, P. V. (2014). Comparative study of bearing materials and failure of plain bearings. *International Journal of Engineering Research & Technology*, *2*(3), 2402–2406.

Demirci, M. T., & Duzcukoglu, H. (2014). Wear behaviors of polytetrafluoroethylene and glass fiber reinforced polyamide 66 journal bearings. *Materials & Design*, *57*, 560–567.

Dhanumalayan, E., & Joshi, G. M. (2018). Performance properties and applications of polytetrafluoroethylene (PTFE)—a review. *Advanced Composites and Hybrid Materials*, *1*(2), 247–268.

Farfan-Cabrera, L. I., & Gallardo-Hernandez, E. A. (2017). Wear evaluation of journal bearings using an adapted micro-scale abrasion tester. *Wear*, *376*, 1841–1848.

Fonte, M., Duarte, P., Anes, V., Freitas, M., & Reis, L. (2015). On the assessment of fatigue life of marine diesel engine crankshafts. *Engineering Failure Analysis*, *56*, 51–57.

Gebretsadik, D. W., Hardell, J., & Prakash, B. (2015). Tribological performance of tin-based overlay plated engine bearing materials. *Tribology International*, *92*, 281–289.

Geng, Z., Hou, S., Shi, G., Duan, D., & Li, S. (2016). Tribological behaviour at various temperatures of WC-Co coatings prepared using different thermal spraying techniques. *Tribology International*, *104*, 36–44.

Ghaziof, S., & Gao, W. (2015). Zn–Ni–Al_2O_3 nano-composite coatings prepared by sol-enhanced electroplating. *Applied Surface Science*, *351*, 869–879.

Goral, A., & Skrzypek, S. J. (2018). The influence of alumina nanoparticles on lattice defects, crystallographic texture and residual stresses in electrodeposited Ni/Al_2O_3 composite coatings. *Applied Surface Science*, *456*, 147–155.

Gu, C., Meng, X., Xie, Y., & Li, P. (2016). A study on the tribological behavior of surface texturing on the nonflat piston ring under mixed lubrication. *Proceedings of the Institution of Mechanical Engineers, Part J: Journal of Engineering Tribology*, *230*(4), 452–471.

Harsimran, S., Santosh, K., & Rakesh, K. (2021). Overview of corrosion and its control: A critical review. *Proceedings of Engineering Sciences*, *3*(1), 13–24.

He, Y., Boluk, Y., Pan, J., Ahniyaz, A., Deltin, T., & Claesson, P. M. (2019). Corrosion protective properties of cellulose nanocrystals reinforced waterborne acrylate-based composite coating. *Corrosion Science*, *155*, 186–194.

Hofsajer, I., & Botef, I. (2013). Cold spray technology for high performance frequency selective conductive structures. *SAIEE Africa Research Journal*, *104*(3), 115–124.

Huang, C., Du, L., & Zhang, W. (2009). Microstructure and tribological properties of plasma sprayed NiCr/Cr_3C_2 and NiCr/Cr_3C_2-BaF_2-CaF_2 composite coatings. *Advanced Tribology*, *1*(1), 669–675.

Jarząbek, D. M., Milczarek, M., Wojciechowski, T., Dziekoński, C., & Chmielewski, M. (2017). The effect of metal coatings on the interfacial bonding strength of ceramics to copper in sintered Cu-SiC composites. *Ceramics International*, *43*(6), 5283–5291.

Jia, Y., Chen, L., Feng, X., Zhou, H., & Chen, J. (2015). Tribological behavior of molybdenum disulfide bonded solid lubricating coatings cured with organosiloxane-modified phosphate binder. *RSC Advances*, *5*(85), 69606–69615.

John, S., Salam, A., Baby, A. M., & Joseph, A. (2019). Corrosion inhibition of mild steel using chitosan/TiO_2 nanocomposite coatings. *Progress in Organic Coatings*, *129*, 254–259.

Kareem, B. (2017). Mechanical failure analysis of automobile crankshafts under service reconditioned modelling approach. *Engineering Failure Analysis*, *80*, 87–101.

Kopeliovich, D. (2011). Geometry and dimensional tolerance of engine bearings. *Engine Professional, AERA*, *1*, 70–76.

Kumar, V., & Balasubramanian, K. (2016). Progress update on failure mechanisms of advanced thermal barrier coatings: A review. *Progress in Organic Coatings*, *90*, 54–82.

Kvryan, A., Carter, N. A., Trivedi, H. K., & Hurley, M. F. (2020). Accelerated testing to investigate corrosion mechanisms of carburized and carbonitrided martensitic stainless steel for aerospace bearings in harsh environments. *Tribology Transactions*, *63*(2), 265–279.

Lee, J. U., Jeong, B., & An, T. H. (2019). Investigation on effective support point of single stern tube bearing for marine propulsion shaft alignment. *Marine Structures*, *64*, 1–17.

Li, B., Li, D., Chen, W., Liu, Y., Zhang, J., Wei, Y., ... & Jia, W. (2019). Effect of current density and deposition time on microstructure and corrosion resistance of Ni-W/TiN nanocomposite coating. *Ceramics International*, *45*(4), 4870–4879.

Ma, C. Y., Zhao, D. Q., Xia, F. F., Xia, H., Williams, T., & Xing, H. Y. (2020). Ultrasonic-assisted electrodeposition of Ni-Al_2O_3 nanocomposites at various ultrasonic powers. *Ceramics International*, *46*(5), 6115–6123.

Mehboob, G., Liu, M. J., Xu, T., Hussain, S., Mehboob, G., & Tahir, A. (2020). A review on failure mechanism of thermal barrier coatings and strategies to extend their lifetime. *Ceramics International*, *46*(7), 8497–8521.

Mehran, Q. M., Fazal, M. A., Bushroa, A. R., & Rubaiee, S. (2018). A critical review on physical vapor deposition coatings applied on different engine components. *Critical Reviews in Solid State and Materials Sciences*, *43*(2), 158–175.

Miguel, M., Leite, M., Ribeiro, A. M. R., Deus, A. M., Reis, L., & Vaz, M. F. (2019). Failure of polymer coated nylon parts produced by additive manufacturing. *Engineering Failure Analysis*, *101*, 485–492.

Moreno-Couranjou, M., Mauchauffe, R., Bonot, S., Detrembleur, C., & Choquet, P. (2018). Anti-biofouling and antibacterial surfaces via a multicomponent coating deposited from an up-scalable atmospheric-pressure plasma-assisted CVD process. *Journal of Materials Chemistry B*, *6*(4), 614–623.

Morris, N., Mohammadpour, M., Rahmani, R., Johns-Rahnejat, P. M., Rahnejat, H., & Dowson, D. (2018). Effect of cylinder deactivation on tribological performance of piston compression ring and connecting rod bearing. *Tribology International*, *120*, 243–254.

Nagata, M., Fujita, M., Yamada, M., & Kitahara, T. (2012). Evaluation of tribological properties of bearing materials for marine diesel engines utilizing acoustic emission technique. *Tribology International*, *46*(1), 183–189.

Niu, Y., Wei, J., & Yu, Z. (2015). Microstructure and tribological behavior of multilayered CrN coating by arc ion plating. *Surface and Coatings Technology*, *275*, 332–340.

Oksanen, V. T., Lehtovaara, A. J., & Kallio, M. H. (2017). Load capacity of lubricated bismuth bronze bimetal bearing under elliptical sliding motion. *Wear*, *388*, 72–80.

Pan, Z., & He, Q. (2015). High cycle fatigue analysis for oil pan of piston aviation kerosene engine. *Engineering Failure Analysis*, *49*, 104–112.

Park, J. H., Kim, H. G., Park, J. Y., Jung, Y. I., Park, D. J., & Koo, Y. H. (2015). High temperature steam-oxidation behavior of arc ion plated Cr coatings for accident tolerant fuel claddings. *Surface and Coatings Technology*, *280*, 256–259.

Pedeferri, P., & Pedeferri, P. (2018). Electrochemical mechanism. *Corrosion Science and Engineering*, *720*, 17–35.

Piazzoni, C., & Milani, P. (2019). The use of spark ablation for generating cluster beams: A review. *Spark Ablation*, *1*(1), 245–271.

Qu, J., Meyer III, H. M., Cai, Z. B., Ma, C., & Luo, H. (2015). Characterization of ZDDP and ionic liquid tribofilms on non-metallic coatings providing insights of tribofilm formation mechanisms. *Wear, 332*, 1273–1285.

Repka, M., Dorr, N., Brenner, J., Gabler, C., McAleese, C., Ishigo, O., & Koshima, M. (2017). Lubricant-surface interactions of polymer-coated engine journal bearings. *Tribology International, 109*, 519–528.

Rudnev, V. S., Vaganov-Vil'kins, A. A., Yarovaya, T. P., & Pavlov, A. D. (2016). Polytetrafluoroethylene-oxide coatings on aluminum alloys. *Surface and Coatings Technology, 307*, 1249–1254.

Santecchia, E., Hamouda, A. M. S., Musharavati, F., Zalnezhad, E., Cabibbo, M., & Spigarelli, S. (2015). Wear resistance investigation of titanium nitride-based coatings. *Ceramics International, 41*(9), 10349–10379.

Scheu, M. N., Tremps, L., Smolka, U., Kolios, A., & Brennan, F. (2019). A systematic failure mode effects and criticality analysis for offshore wind turbine systems towards integrated condition-based maintenance strategies. *Ocean Engineering, 176*, 118–133.

Shakhova, I., Mironov, E., Azarmi, F., & Safonov, A. (2017). Thermo-electrical properties of the alumina coatings deposited by different thermal spraying technologies. *Ceramics International, 43*(17), 15392–15401.

Sharma, V. K., Singh, R. C., & Chaudhary, R. (2019). *Experimental Study of Sliding Wear Behavior of the Casted Lead Bronze Journal Bearing Material* (No. 2019-01-0824). SAE Technical Paper.

Shirazi, M. G., Rashid, A. S. B. A., Nazir, R. B., Rashid, A. H. B. A., Moayedi, H., Horpibulsuk, S., & Samingthong, W. (2020). Sustainable soil bearing capacity improvement using natural limited life geotextile reinforcement—a review. *Minerals, 10*(5), 479.

Summer, F., Grun, F., Offenbecher, M., & Taylor, S. (2019). Challenges of friction reduction of engine plain bearings—Tackling the problem with novel bearing materials. *Tribology International, 131*, 238–250.

Szubert, K., Dutkiewicz, A., Dutkiewicz, M., & Maciejewski, H. (2019). Wood protective coatings based on fluorocarbosilane. *Cellulose, 26*(18), 9853–9861.

Unal, E., & Karahan, I. H. (2018). Effects of ultrasonic agitation prior to deposition and additives in the bath on electrodeposited Ni-B/hBN composite coatings. *Journal of Alloys and Compounds, 763*, 329–341.

Vetter, J., Barbezat, G., Crummenauer, J., & Avissar, J. (2005). Surface treatment selections for automotive applications. *Surface and Coatings Technology, 200*(5-6), 1962–1968.

Walsh, F. C., & Low, C. T. J. (2016). A review of developments in the electrodeposition of tin. *Surface and Coatings Technology, 288*, 79–94.

Wan, H. P., & Ni, Y. Q. (2018). Bayesian modeling approach for forecast of structural stress response using structural health monitoring data. *Journal of Structural Engineering, 144*(9), 04018130.

Witek, L., Sikora, M., Stachowicz, F., & Trzepiecinski, T. (2017). Stress and failure analysis of the crankshaft of diesel engine. *Engineering Failure Analysis, 82*, 703–712.

Wu, L., Fu, Y., Li, M., Liang, G., Zhang, Y., & Cui, Y. (2021). The effects of material and structure of main bearing caps on crankshaft lubrication of diesel engine. *International Journal of Engine Research, 22*(4), 1086–1100.

Yang, Y., Zhou, J., Detsch, R., Taccardi, N., Heise, S., Virtanen, S., & Boccaccini, A. R. (2018). Biodegradable nanostructures: Degradation process and biocompatibility of iron oxide nanostructured arrays. *Materials Science and Engineering: C, 85*, 203–213.

Zhang, W., Mei, T., Li, B., Yang, L., Du, S., Miao, Y., & Chu, H. (2021). Effect of current density and agitation modes on the structural and corrosion behavior of Ni/diamond composite coatings. *Journal of Materials Research and Technology*, *12*, 1473–1485.

Zorn, K., Eberhard, A., Schinagl, M., & Bakk, E. (2020). Lead-free bearings for high performance gas engines. *MTZ Worldwide*, *81*(5), 58–63.

4 Corrosion
A Science of Degradation

4.1 CORROSION

Corrosion is a natural phenomenon; it is defined as the physiochemical activities between the metal and the surrounding environment leading to material degradation and thus resulting in impairment and lowering of the integrity of the structure. This natural reaction can be a general corrosion process or via consistent loss of metal on the visible surface and sometimes may be localized corrosion where only a limited portion of the surface is in contact with the environment (Fingsgar & Jackson, 2014). The appearances of material degradations are often traced to the varying conditions and changes in fluid compositions, changes in operating conditions of the pressures and temperatures, the underlying or contact materials with working metals, and other atmospheric conditions. The degree of corrosion reactivity depends upon the sensitivity of a particular metal or alloy to a specific medium or environment (Walton et al., 2014). It may, therefore, be observed that corrosion is a potent force which destroys the economy, depletes resources, and causes costly and untimely failures of plants, equipment, and components. In general terms, environmental influences contribute immensely to the corrosion spontaneous reaction of metals. The environment consists of the entire surroundings in contact with the material. The main factors that define the environment are: (a) physical state—gas, liquid, or solid; (b) chemical composition—constituents and concentrations; and (c) temperature. Other factors can be important in specific cases: examples of these factors are the relative velocity of a solution (because of flow or agitation) and mechanical loads on the material, including residual stress within the material (Fingsgar & Jackson, 2014). Corrosion failure studies have increasingly become important in science and engineering research. In an attempt to prevent failure reoccurrence, the need to identify the form of corrosion problem and its mechanism is needful. The following are some of the various types in which corrosion occurs.

4.1.1 Uniform Corrosion

Uniform corrosion by definition is an attack characterized by an even distribution of corrosion occurrence over the surface area. This consequently leads to a uniform

DOI: 10.1201/9781003562078-4

thickness reduction. In many cases, uniform corrosion is relatively measured and foreseen, making disastrous failures relatively uncommon. Homogeneous materials without a substantial passivation tendency in the actual medium are predisposed to this form of corrosion. Uniform corrosion is presumed as one of the most common forms of corrosion and is particularly responsible for most of the material loss (Walton et al., 2014). Surface corrosion can indicate a breakdown in the protective coating system; however, it should be examined closely for more advanced attack. If surface corrosion is permitted to continue, the surface may become rough, and surface corrosion can lead to more serious types of corrosion.

4.1.2 Galvanic Corrosion

Galvanic corrosion occurs when dissimilar metallic materials are brought into contact in the presence of an electrolyte. Such damage can also occur between metals and alloys and other conducting materials such as carbon or graphite. An electrochemical corrosion cell is set up due to differences in the corrosion potentials of dissimilar materials. The material with the more noble corrosion potential then becomes the cathode of the corrosion cell, whereas the less noble material is consumed by anodic dissolution (Roy et al., 2014).

4.1.3 Thermogalvanic Corrosion

Temperature induced on material in a corrosive environment results in thermogalvanic corrosion. Usually, the hot surface forms the anode and the cold one the cathode. Partly, this is so because the anodic properties of the material depend on the temperature. In addition, the properties of the environment along the metal surface will also vary due to varying temperatures and varying temperature gradients normal to the metal surface. Often this affects the cathodic reaction, which should also be taken into consideration when such a corrosion form is analysed. Thermogalvanic corrosion is prevented by appropriate design and measures to avoid uneven heating/cooling and forming of hot spots (Fernández-Domene et al., 2014).

4.1.4 Crevice Corrosion

Crevice corrosion often occurs in an area in which the gap is sufficiently wide for a liquid to penetrate into the crevice and sufficiently narrow for the liquid in the crevice to be stagnant. Because oxygen diffusion into the crevice is restricted, a differential aeration cell tends to be set up between the crevice (microenvironment) and the external surface. The cathodic oxygen reduction reaction cannot be sustained in the crevice area, giving it an anodic character in the concentration cell. This anodic imbalance can lead to the creation of highly corrosive micro-environmental conditions in the crevice, conducive to further metal dissolution. The corrosion progress appears beneath flange gaskets, nail and screw heads, and paint coating edges, in overlap joints, between tubes and tube plates in heat exchangers, and so on (Han et al., 2014).

4.1.5 Pitting Corrosion

Pitting corrosion is a localized form of corrosion by which holes are produced in the material. These kinds of corrosion are considered to be more dangerous than general corrosion damage because of the difficulty of detecting and predicting in a design system. Pitting corrosion occurs on more or less passivated metals and alloys in environments containing chloride, bromide, iodide, or perchlorate ions when the electrode potential exceeds a critical value, the pitting potential, which depends on various conditions. Pitting is a dangerous form of corrosion since the material in many cases may be penetrated without a prior notice of pit sign and growth (Li et al., 2014).

4.1.6 Intergranular Corrosion

Intergranular corrosion is a localized form of corrosion attack on or at grain boundaries with insignificant corrosion on other parts of the surface. The attacks propagate into the material (Wang et al., 2013). This is a dangerous form of corrosion because the cohesive forces between the grains may be too small to withstand tensile stresses; the toughness of the material is seriously reduced at a relatively early stage, and fracture can occur without warning. Grains may fall out, leaving pits or grooves, but this may not be particularly important. The general cause of intergranular corrosion is the presence of galvanic elements due to differences in the concentration of impurities or alloying elements between the material in or at the grain boundaries and the interior of the grains (Han et al., 2014).

4.1.7 Hydrogen Embrittlement

In general, embrittlement corrosion is corrosion that causes a ductile material to fail without localized yielding or shearing. More specifically, hydrogen embrittlement assumes several different forms with a general similarity. This damage takes place at the cathode, an area that we normally think is safe from corrosion remember the saying that corrosion takes place at the anode, but it does not in this case. Hydrogen ions are reduced to hydrogen molecules at the cathode. Those atoms usually pair up to become hydrogen molecules. These molecules harmlessly bubble off as hydrogen gas (Frankel et al., 2013).

4.1.8 Erosion and Abrasion Corrosion

When there is a relative movement between a corrosive fluid and a metallic material immersed in it, the material surface is in many cases exposed to mechanical wear effects leading to increased corrosion, which we usually call erosion corrosion. The mechanism is that deposits of corrosion products, or salts precipitated because of the corrosion process, are worn off, dissolved, or prevented from being formed so that the material surface becomes metallically clean and therefore more active (Lu et al. 2014). In extreme cases, erosion corrosion may be accompanied by pure mechanical erosion, by which solid particles in the fluid may tear out particles from the material itself and cause plastic deformation, which may make the metal even more active.

The results of erosion corrosion are grooves or pits with a pattern determined by the flow direction and the local flow conditions. Reasonably, the corrosion form is typical at relatively high velocities between the material surface and the fluid, and it is particularly intensive in cases of two-phase or multi-phase flow (Walton et al., 2014).

4.2 CORROSION PROTECTION

The challenge faced by metal due to corrosion activities has made coatings a household protective measure for failure prevention. In the past years, coatings applied using various techniques were done to improve the performance and lifetime of components for properties such as chemical stability, wear resistance, and hardness performance. Protective coatings can be broadly divided into the following.

4.2.1 Metallic Coatings

Metallic coatings involve the deposition of novel metal coatings by an electroplating process, electroless plating, spraying, hot dipping, CVD, and ion vapour deposition. Some important materials for coatings are cadmium, chromium, nickel, aluminium, and zinc. They are used in applications requiring significant reflection and strong adhesion of metallic materials. Metallic coatings are known to show improved performance with lower cure temperature and significant corrosion resistance (Tan et al., 2008).

4.2.2 Inhibitive Coatings

In most advanced coatings where corrosion resistances are paramount, inhibitors are added to the bath to act as a barrier to the environment. Such inhibitor plays an active part in the curing, adhesion, and surface appearance of the produced alloy. Inhibitors such as cinnamic acid are added to the paint bath to prevent degradation of steel in neutral and acidified media (Tuaweri & Wilcox, 2006).

4.2.3 Inorganic Coatings

Inorganic coatings involve ceramic and glass fabricated through chemical action, with or without electrical support. They are usually impervious to water solution. Sometimes these coatings are applied as pre-coating before the co-deposition proper, i.e., treatments can also be a preparatory step prior to painting. The treatments change the immediate surface layer of metal into a film of metallic oxide or compound that has better corrosion resistance than the natural oxide film and provides an effective protection such as anodizing (Lakshmi et al., 2014).

4.2.4 Organic Coatings

The development of organic coating is to reduce the environmental distortion that often occurs in carriage vessels in industries. Paints and high-performance organic coatings were developed to protect equipment from these challenges, especially in petrochemical settings. Obviously, the petroleum cracking product dislodge often contains unsaturated workable compounds. Epoxy, polyurethane, chlorinated rubber, and polyvinyl chloride coatings are extensively used today in the manufacturing industry to attack water and oxygen and prevent the occurrence of a cathodic reaction beneath the coating. The barrier properties are further increased by the addition of inhibitors, such as chromate in the primer (Yang et al., 2007).

4.2.5 Electrodeposition

Deposition practices are conceded in order to obtain modified surface properties over based metal through the addition of a layer of another metal alloy, composite, or by the formation of an oxide film from an electrolyte (Popoola et al., 2012). Interestingly the goal of the industry in co-deposition technique is to enhance the excellent value of metal through appearance improvement, and better mechanical and electrochemical resistance characteristics (Shanaghi et al., 2009). However, surface behaviour tailored towards advanced surface improvement will impart corrosion resistance and subsequently physical or mechanical characteristics usually by wear and tribological resistance, hardness and thermal stability. Ordinarily, different methods have been used to obtain good surface finishing for metal deposition but the electroplating route provides a sustainable route to achieve enhanced characteristics (Rusu et al., 2012). Another feature of an electroplating technique is to give a homogeneous deposit with strong adhesion to the working substrate. The kinds of coating properties are often tailored to the required specifications and properties intended to be achieved. Moreover, many metals may be deposited with different properties based on bath formation and deposition conditions. It is for these reasons that the condition for electroplating differs for each metal, ranging from the bath composition, current density and temperature, conducting salts, the anode, and the pH (Fenker et al., 2014). With excellent process parameters, the deposition baths have a tendency for stability over a period of time and pose characteristics for reproducibility. In general, for an effective plating process, the precise pretreatment of the cathode and careful anode selection, plating bath, current density, and other electrolysis conditions are essential factors for excellent coating properties.

4.2.6 Selective Coating

In this process, electro-deposition can be made on the desired localized areas without the need for masking and without immersion of components. The anode is mounted in an insulated handle and covered by an absorbent pad soaked in the electrolyte. The work is connected to the negative side of a DC power source and the circuit is completed by the contact of an absorbent pad with the workpiece. The process makes masking unnecessary. The deposition rate is higher than vat plating and this process enjoys the benefit of portability.

4.2.7 Electroless Plating

In electroless plating, the process involves an autocatalytic reduction reaction of metallic ions in an aqueous solution and the subsequent deposition of cathode metal without the use of an electrical energy source. The electroless plating does not use electrodes, rather the process uses an immersion plating solution with a displacement of the surface skin of the cathode substrate by a novel noble metal that is in solution. Electroless has outstanding advantages, which include better deposit quality, and improved physical and mechanical properties. The electroless deposit often gives distinct advantages when plating irregularly shaped objects, holes, recesses, internal surfaces, valves, threaded parts, and so forth. In all, the most electroless process uses variables that are desirable for better performance such as pH, concentration of the bath, and temperature. For perfect electroless deposition, a bath must contain a source of ions, a reducing agent, a stabilizer, a buffering agent, a complexing agent, and a wetting agent (Sudagar et al., 2013).

4.2.8 Flame Spraying

In flame spraying, the aim is to melt the coating material and blow it onto the surface to be coated. The coating material is in the form of very small molten particles or droplets. Four methods based on the form of the coating material are generally used: an electric arc (Yadov et al., 2006). The resulting molten metal is blown out of the arc by an auxiliary gas stream, as droplets. The process is versatile with low capital investment. In the electrostatic spraying process, the particles released from the spray gun are electrostatically charged and propelled at low velocity by air or a revolving spray head. Too much air pressure is to be avoided (Nelson et al., 2011).

4.2.9 Pack Cementation

Pack cementation techniques are well-known to convene oxidation resistance on ferrous alloys. They are relatively expensive. They are also known as diffusive coatings and they exhibit a compositionally graded layer with the provision for a strong bonding between the coating materials and based substrate. The coating process is cost-effective and mass production can be done even for complex shapes compared to other coating routes like plasma and cold spraying coatings. Pack cementation processes include aluminizing, chromizing, and siliconizing. Components are packed in metal powders in sealed heat-resistant retorts and heated inside a furnace to precisely control temperature–time profiles (Park et al., 2008).

4.2.10 Hot-Dip Galvanizing

The hot-dip galvanizing process is a distinctive coating technique with a significant advantage over other protective measures, especially for the corrosion protection route. They are utilized in major fabrications and large-scale production in the iron and steel industry. Galvanizing can be found in almost every major application and industry where iron or mild steel is used. The technique provides electrochemical protection for steel by the zinc coatings galvanizing process. In fact, all produced components rely on the cathodic protection provided by zinc to prevent corrosion of exposed steel at cut edges (Li et al., 2010).

4.2.11 Vacuum Deposition

In vacuum deposition, the effective coating process is determined by the effective optimization of the process parameters such as vapour pressure, atomic weight, the metallizing concentration of the deposited particulate, and the evaporated temperature through a current passage. The operation chamber is maintained in a high vacuum. The surface is often at room temperature or relatively lower temperature. The vaporized aluminium is condensed on the surface of the workpiece. Coatings of the order of 0.021–0.075 mm are generally applied (Mukherjee & Barhai, 2012).

4.2.12 Vapour Deposition

Vapour deposition is a gas process where reactant precursor gases form a solid coating on a heated substrate in a reaction chamber. In this process, vapours of a metal-bearing compound are brought into contact with a heated substrate and a metal compound is deposited on the surface. Steel is exposed to dry aluminium chloride in a reducing atmosphere at 1000°C and metallic particulate is deposited on the surface. The properties often obtained from vapour deposition include high resistance to abrasive and adhesion wear, fine grains, and high purity (Tsubakino et al., 2013).

REFERENCES

Fenker, M., Balzer, M., & Kappl, H. (2014). Corrosion protection with hard coatings on steel: Past approaches and current research efforts. *Surface & Coatings Technology*, *257*, 182–205.

Fernández-Domene, R. M., Blasco-Tamarit, E., García-García, D. M., & Garcia-Anton, J. (2012). Thermogalvanic corrosion of Alloy 31 in different heavy brine LiBr solutions. *Corrosion Science*, *55*, 40-53.

Fingsgar, M., & Jackson, J. (2014). Application of corrosion inhibitors for steels in acidic media for the oil and gas industry: A review. *Corrosion Science*, *86*, 17–41.

Frankel, G. S., Samaniego, A., & Birbilis, N. (2013). Evolution of hydrogen at dissolving magnesium surfaces. *Corrosion Science*, *70*, 104–111.

Han, W., Pan, C., Wang, Z., & Yu, G. (2014). A study on the initial corrosion behavior of carbon steel exposed to outdoor wet-dry cyclic condition. *Corrosion Science*, *88*, 89–100.

Lakshmi, R. V., Yoganandan, G., Mohan, A. V. N., & Basu, B. J. (2014). Effect of surface pre-treatment by silanization on corrosion protection of AA2024-T3 alloy by sol–gel nanocomposite coatings. *Surface & Coatings Technology*, *240*, 353–360.

Li, J. K., Ma, H. X., Zhu, S. D., Qu, C. T., & Yin, Z. F. (2014). Erosion resistance of CO_2 corrosion scales formed on Api P110 carbon steel. *Corrosion Science*, *86*, 101–107.

Li, M. C., Xin, S. S., & Wu, M. Y. (2010). Electrodeposition behaviour of Mg with Mg with Zn from acidic sulphate solutions. *Journal of Solid-State Electrochemistry*, *14*, 2235–2240.

Lu, J. Z., Qi, H., Luo, K. Y., Luo, M., & Cheng, X. N. (2014). Corrosion behaviour of AISI 304 stainless steel subjected to massive laser shock peening impacts with different pulse energies. *Corrosion Science*, *80*, 53–59.

Mukherjee, S. K., & Barhai, P. K. (2012). Reliability of anodic vacuum arc in depiniting thermoelectric alloy thin films. *Journal of Alloys and Compounds*, *511*, 14–21.

Nelson, G. M., Nychka, J. A., & Mcdonald, A. G. (2011). Flame spray deposition of titanium alloy-bioactive glass composite coatings. *Journal of Thermal Spray Technology*, *20*, 1339–1351.

Park, J. S., Kim, J. M., Kim, H. Y., Lee, J. S., Oh, I. H., & Kang, C. S. (2008). Surface protection effect of diffusion pack cementation process by Al-Si powders with chloride activator on magnesium and its alloys. *Materials Transactions*, *49*, 1048–1051.

Popoola, A. P. I., Fayomi, O. S. I., & Popoola, O. M. (2012). Comparative study of microstructure, tribological and corrosion properties of plated Zn and Zn alloy coating. *International Journal of Electrochemical Science*, *7*, 4860–4870.

Roy, P., Karfa, P., Adhikari, U., & Sukul, D. (2014). Corrosion inhibition of mild steel in acidic medium by polyacrylamide grafted guar gum with various grafting percentage: Effect of intramolecular synergism. *Corrosion Science*, *88*, 246–253.

Rusu, D. E., Ispas, A., Bund, A., Gheorghies, C., & Cârâc, G. (2012). Corrosion tests of nickel coatings prepared from a Watts-type bath. *Journal of Coating Technology Resources*, *9*(1), 87–95.

Shanaghi, A., Sabour, A. R., Shahrabi, T., & Aliofkhazraee, M. (2009). Corrosion protection of mild steel by applying TiO_2 nanoparticle coating via sol-gel method. *Protection of Metals and Physical Chemistry of Surfaces*, *45*, 305–311

Sudagar, J., Lian, J., & Sha, W. (2013). Electroless nickel, alloy composite and nano coatings. *Journal of Alloys and Compounds*, *571*, 183–204.

Tan, C., Liy, Y., Zhao, X., & Zheng, Z. (2008). Nickel co-deposition with SiC particles at initial stage. *Transactional Non-Ferrous Metals Society of China*, *18*(5), 128–1133.

Tsubakino, D., Nomura, K., & Yamashit, Y. (2013). Boundary control of boundary coupled parabolic distributed parameter systems on partial state transformation, *26*(7), 277–287.

Tuaweri, T. J., & Wilcox, G. D. (2006). Behavior of Zn-SiO_2 electro deposition in the presence of N, N-dimethyldodecylamine. *Surface and Coating Technology. Transactions of the Institute of Systems, Control and Information Engineers*, *200*, 5921–5930.

Walton, C. A., Martin, H. J., Horstemeyer, M. M. F., Whittington, W. R., Horstemeyer, C. J., & Wang, P. T. (2014). Corrosion stress relaxation and tensile strength effects in an extruded Az31 magnesium alloy. *Corrosion Science*, *80*, 503–510.

Wang, X., Wan, Y., Wang, Q., & Ma, Y. (2013). Study of inhibitive effect of 1, 4-bis (benzimidazolyl) benzene on mild steel corrosion in 0.5 M HCl and 0.25 M H_2So_4 solutions. *International Journal of Electrochemical Science*, *8*, 806–820.

Yadov, A. K., Arora, N., & Dwivedi, D. K. (2006). On microstructure, hardness and wear behaviour of flame sprayed Co base alloy coating deposited on mild steel, *Surface Engineering*, *22*, 331–336.

Yang, Y. J., Chen, J. G., Chi, F. P., Zhu, Q. R., Zhang, H. X., & Sun, J. Z. (2007). Ultra-high harmonic generation from atom with superposition of ground highly exited state. *Chinese Physics Letter*, *24*, 1537–1540.

5 Deposition of Binary and Quaternary Alloys on Steel for Maximal Performance Use

5.1 MILD STEEL

Mild steel is by far the most widely used and one of the world's cheapest and most useful metals. It is essentially alloys of iron and carbon with small additions of elements such as manganese and silicon added to provide the requisite mechanical properties (Derek et al., 2024). The properties of mild steel depend primarily on the amount of carbon it contains (Ranganatha et al., 2012). Mild steel contains carbon of up to 0.25% in constituent among other elements. This made it useful to be turned into a wide range of products, including structural beams, car bodies, kitchen appliances, and cans treatment (Jiang et al., 2012). Considering the microstructure of mild steel for instance with 0.2% carbon, such steel consists of about 75% of proeutectoid ferrite that forms above the eutectoid temperature and about 25% of pearlite (pearlite and ferrite being microstructure components of steel). When the carbon content in the steel is increased, the amount of pearlite increases until we get the fully pearlitic structure of a composition of 0.8% carbon (Majumdar et al., 2010). However, in slowly cooled mild steels, the overall hardness and ductility of the steel are determined by the relative proportions of the soft, ductile ferrite and the hard, brittle cementite. Hence, so as to be able to respond to such a great demand and to suit the requirements of different applications, mild steel needs to offer several desired properties and these properties are achieved by alloying it (Abdullah et al., 2008).

A study of the constitution and structure of all steels and irons must first start with the iron–carbon equilibrium diagram. The iron–carbon diagram provides a valuable foundation on which to build knowledge of steel transformation. Many of the basic features of this system influence the behaviour of even the most complex alloy steels. For example, the phases found in the simple binary Fe–C system of mild steel persist in complex steels, but it is necessary to examine the effects that alloying elements have on the formation and properties of these phases (Panossian et al., 2012). It should be first pointed out that the normal equilibrium diagram really represents the metastable equilibrium between iron and iron carbide (cementite). Cementite is metastable, and the true equilibrium should be between iron and graphite. Although graphite occurs extensively in cast irons (2–4 wt%C), it is usually difficult to obtain this equilibrium phase in mild steel (0.03–1.5 wt%C) due to its homogeneity (Guo et al., 2012). The much larger phase field of γ-iron (austenite) compared with that of

DOI: 10.1201/9781003562078-5

α-iron (ferrite) reflects the much greater solubility of carbon in γ-iron, with a maximum value of just over 2 wt% at 1147°C. This high solubility of carbon in γ-iron is of extreme importance in heat treatment, when solution treatment in the γ-region followed by rapid quenching to room temperature allows a supersaturated solid solution of carbon in iron to be formed (Ehteram & Aisha, 2008).

The α-iron phase field is severely restricted, with a maximum carbon solubility of 0.02 wt% at 723°C (P), so over the carbon range encountered in steels from 0.05 to 1.5 wt%, α-iron is normally associated with iron carbide in one form or another (Han et al., 2014). Similarly, the δ-phase field is very restricted between 1390°C and 1534°C and disappears completely when the carbon content reaches 0.5 wt% (B). There are several temperatures or critical points in the diagram, which are important, both from the basic and the practical point of view (Liu et al., 2014). Furthermore, it has been seen that hardness, brittleness, and ductility are very important properties as they determine mainly the way these different metal steels are used. During processing and forging, the mild steel surface is oxidized by air, and the scale produced is usually termed millscale. In air, the presence of millscale on the mild steel may reduce the corrosion rate over comparatively short periods, but over longer periods the rate tends to rise. In water, severe pitting of the steel may occur if large amounts of millscale are present on the surface. For example, in the open air, a mild steel containing 0.2% Si rusts about 10% less rapidly than an otherwise similar steel containing 0.02% Si (Roy et al., 2014).

In reality, mild steel is selected not for its corrosion resistance but for such properties as strength, ease of fabrication, and cost. Primarily mild steel has quite good resistance to alkalis, many organics, and strong oxidizing acids. As a general rule, acids should be avoided. Mild steel can be susceptible to SCC in media that contain nitrates, hydroxides, ammonia, and hydrogen sulide. Any evolved hydrogen may cause embrittlement and blistering in the steel (Melchers & Chervon, 2010). Like other metals that form passive oxide films, iron benefits in situations in which there is essentially no oxygen to depolarize the cathodic reaction or sufficient oxidizing power to form a stable oxide film. The corrosion rates of mild steel when immersed in seawater or buried in soil are not the same as other metals. Mild steel has a number of phases and homogeneities at the surface, which can cause local cells. The corrosion resistance of iron is low because cathodic reduction can easily take place on its surface; moreover, its corrosion product is porous and non-adherent. Mild steel finds extensive application primarily because of its low cost, reasonably good mechanical properties, and ease of fabrication. Ambient conditions in an industrial environment are relatively more corrosive because of the presence of moisture and chemical pollutants in the air. Chlorides in coastal areas and sulphur dioxide are highly aggressive, and they lower the critical humidity level for the onset of corrosion (Olivaries-Xomelt et al., 2013). Despite these shortcomings, mild steels, with or without minor alloying elements, are widely used as the most economical materials for construction under ambient, aggressive conditions, and with various combinations of protective coatings and other corrosion prevention or control methods (Fingsgar & Jackson, 2014). Corrosion of mild steel in the presence of moisture is particularly severe when acidic gases or vapours are involved, such as oxides of nitrogen and sulphur or chlorine. At high temperatures, water vapour does not contribute to corrosion, but once the dew point is reached, condensation takes place, and the corrosion rate rises drastically (Soares et al., 2009).

5.2 METALLIC ELEMENTS IN STEEL PROTECTIVE COATING

5.2.1 Zinc

Zinc has long been in use as a sacrificial anode to protect steels either by electrodeposition or hot-dipping route. The steel protections by zinc are used in automobiles, electric appliances, automobiles, and many more. Zinc coating sometimes gives up to 25 μm thickness and may be as thin and less on threaded parts. Electrodeposited zinc gives thinner coatings as compared to the hot dipping process. The limited life span of zinc coating systems in industrial atmospheres was attributed to the attack by sulphuric acid in polluted atmospheres and aggressive nature of some environments. Zinc plating does not perform as well as zinc alloy in tropical and marine atmospheres due to the promising effect of the inert particulate (Gomes, Frade et al., 2012).

5.2.2 Aluminium

Aluminium is a non-magnetic and non-sparking silvery-white material and a member of the boron family. After oxygen and silicon, it remains the most abundant metal in the Earth's crust. It is soft, durable, and has lightweight characteristics. Corrosion resistances of aluminium alloys are excellent due to a thin surface layer of aluminium oxide that forms when the metal is exposed to air, effectively simultaneously preventing further oxidation. Aluminium coatings can be applied to steel by hot dipping, cementation, ion vapour deposition, and spraying. Ion vapour deposition is a relatively new process, and spraying is the only process that has been used extensively over a long period of time. Sprayed aluminium coatings provide an adherent, somewhat absorbent film about 100–150 μm thick. They provide very good protection to steel, and they may be sealed with organic lacquers or paints to provide further protection and delay the formation of visible surface rust over the substrate (Turk et al., 2007).

5.2.3 Titanium

Titanium and titanium IV oxide have conventionally been used in aerospace, medical, automotive, and sporting goods applications due to their high strength and stiffness-to-weight ratio and excellent biocompatibility. Currently, titanium alloys have a diversity of place applications in these industries, but the increased use of titanium IV oxide has been linked to its relatively stability in aqueous solution and moderately inexpensive. There is continued interest in increasing the use of titanium alloys by developing lower-cost processes, improved properties, improved biocompatibility, and higher buy-to-fly ratios. Titanium particles are interesting because of their semiconductor properties with vast application in photo-catalysts and high temperatures. Titanium and its oxide are non-toxic; they possess good thermomechanical processing properties (Praveen & Venkatesha, 2008).

5.2.4 Tin

Tin is a soft, pliable, silvery-white metal. It has a low oxidizing potential and resists corrosion because it is protected by an oxide film. Tin is indispensable due to its exceptional characteristics in the production of multi-functional materials in mechanical and electrical uses such as electric capacitors and components of integrated circuits. Tin is non-toxic, and unreactive to water and oxygen, as it is protected by an oxide film (Finazzi et al., 2004). It dissolves in acids and bases. When heated in air tin forms tin (IV) oxide, which is feebly acidic and assists in enhanced application for conductive coatings glass. Tin characteristics are referred to as good players in the development of ceramic appearance. Their oxides are variegated with other metals to provide unique characteristics for engineering applications (Hu & Wang, 2006). Tin originated in group IV of the Periodic Table with carbon, silicon, germanium, and lead. It displays the oxidation states +2 and +4. In the +2 state, it is basic and behaves like a metal. In its +4 state, it is amphoteric and can behave in an acidic manner in an alkaline solution. Tin (II) chloride (stannous chloride) is a reducing agent, while tin (IV) chloride (stannic chloride) is amphoteric. Tin and its oxide are mostly used in alloys, and in tin plate, which is a thin sheet of steel with a protective coating of tin. Tin plate is used for food cans because it is not reactive to the acids present in food. Alloys of tin include bronze (a combination of tin and copper), pewter (a combination of tin and lead), superconducting wire (a combination of tin and niobium), Babbitt metal (a combination of tin, copper and antimony), and bell metal (a combination of tin and copper and solder which is a combination of tin and lead). Babbitt metal is used for the surface of bearings. Superconducting wires are used in the manufacture of extremely powerful magnets (Arici et al., 2011).

5.2.5 Co-deposited Metallic Materials

The major advancement in metallic materials is the use of combined metallic alloy to produce sustains protection in both chemical and mechanical resistance applications. For instance, in aerospace applications, chromium combines with aluminium giving anti-corrosion and wear resistance with activeness in painting and decoration. More so copper–nickel formation both in electroless and electrodeposition techniques enhances the life span propagation of the cathode and gives low internal stress. Zinc and magnesium coatings, which are often produced through conventional processes, are observed to provide densely packed corrosion products which enhance corrosion resistance (M. C. Li et al., 2010). Tuaweri and Wilcox (2006) attested that suspended inert particles such as silica with zinc offer unique functional properties like corrosion improvement, wear and hardness resistance, lubrication, and superconductive characteristics. More so, this increase in hardness is achieved by a precipitation hardening process involving phosphorous, chromium, and cadmium, which is usually present in amounts of 7–15% for thermo-mechanical and tribo-oxidation properties. For superalloy stability, the heat treatment is observed at temperatures of about 200°C–600°C (Fayomi et al., 2014). Conclusively, zinc and aluminium, zinc and zirconia, zinc and titanium, and so on will possibly yield good co-deposition, reduced mechanical failure, and sustained corrosion improvement if proper variable process parameters are followed. This is not because the electrodeposited alloys are not desirable, but because the closer control required for the alloy deposition complicates the commercial electroplating technique (Xia et al., 2009).

5.3 STEEL'S CHARACTERISTIC RESPONSES UPON DEPOSITION

Corrosion, thermal instability, and wear failures are the most frequently encountered failures in major mechanical and chemical components. According to Montemor, (2004), when these catastrophes exist concurrently together their existence often leads to a more severe humiliation of metals. Mild steels are widely used as structural components in automobile, petrochemical, and marine industries due to their good mechanical properties (Sancakoglu et al., 2011). However, the author's submissions are restricted owing to poor mechanical and chemical characteristics. According to Tiwari et al. (2011), "mild steel possesses reasonably good strength but exhibit poor resistance to corrosion". When mild steel comes in contact with alkalis and acidified media such as HCl and NaCl they are susceptible to corrosion degradation which imposes a major limitation on its usage in severe atmospheric environments. The authors indicated that increasing the corrosion resistance of mild steel requires the formulation of a suitable conversion coating to provide an adherent and protective layer on a mild steel substrate. An adaptation coating was formulated and applied on mild steel prior to sol–gel Al_2O_3 by using silica sol and aluminium oxyhydroxide followed by heat treatment at 500°C. The transformation coating formed a composite oxide containing Al_2O_3, SiO_2, α-Fe_2O_3, and Fe_3O_4 on the surface of the substrate. From the examined study, the results showed barrier-type protection of the substrate with improved corrosion resistance of mild steel. The sol–gel coating reduced the corrosion current density of the substrate by five orders of magnitude with improved protection over the substrate.

According to Panagopoulos et al. (2011), "Poor tribological and plastic deformation of mild steel is a huge concern of its usage in advance manufacturing application". The authors attested that abrasive wear components such as metal tends to deform mild steel surfaces severely during abrasive operation. These deformation challenges attested by numerous authors on mild steels inflict their disadvantages. The authors affirmed that the efficient approach to improve tribological and plastic deformation is to fabricate a hard coating alloy on mild steel. A developed composite multi-layered coating, which consisted of an electrodeposited Zn–Fe alloy layer, a zinc phosphate conversion layer, and one, two, or three organic layers, were deposited on a mild steel substrate. The adhesion between the multi-layered coating and the mild steel substrate was studied with the aid of a scratch-testing technique. Observation of the worn surface of different multi-layered coatings was performed with the aid of metallurgical microscopy. The same multi-layered coatings were examined with Fourier transform infrared (FTIR) spectroscopy and XRD techniques. From the studies, experimental deductions indicated that multi-layered coatings appeared to be continuous, uniform in thickness, and without cracks or flaws. The multi-layered coating with organic layers had the highest cohesion and adhesion strength compared to the steel substrate. All multi-layered coatings increased the tribological resistance of mild steel substrate.

According to Mahieu et al. (1999), the cost incurred due to the failure, and repeated damage of mild steel in service conditions, is a huge detriment to the integrity of steel usage in hash conditions. The possibility of increasing the corrosion resistance and cost of mild steel has been increasing paths since it is relatively cheap and possesses good weldability. The authors resolve that electrodepositing becomes a preferred method depending on the admixed alloy particle and the process variables. With proper electrodeposition parameters such as current density, electrolyte flow rate, and pH, deposited steel will provide high chemical and mechanical resistance to deformation and hence reduce cost wastage. The authors elucidate this fact with cobalt contents varying from 0.2 to 7 wt% in zinc electrolyte. The influence of these process parameters on the characteristics of the coating morphology could be related to the hydroxide suppression mechanism for anomalous co-deposition (ACD). The application tests showed that zinc–cobalt alloy coatings have a good adhesion level when compared to pure zinc coatings and base mild steel. On the other hand, the corrosion resistance of a homogeneous film-forming zinc–cobalt alloy coating is significantly better than that of the substrate, which will apparently reduce cost.

Dong et al. (2009) stated that the anticipated performance of mild steel under service conditions that require thermal stability and improved microstructural properties has fallen out because of low thermal resistance in a high-temperature environment. In view of the demands and application of mild steel in harsh industrial usage, several techniques have been employed with attractive mechanical behaviour for high-temperature conditions but with poor oxidation instability. With the intention of improving the desired thermal stability, the author studies the thermal properties of dispersible SiO_2 nanocomposite in Ni/P electrolyte on steel substrate in the absence of surfactant. The resulting Ni/P-SiO_2 composite coating was heat-treated for one hour at 200°C, 400°C, and 600°C, respectively. More so, the structural changes of the composite coatings before and after heat treatment were measured and investigated. Results show that co-deposited SiO_2 particles contributed to improved thermal stability and the composite coating heat-treated at about 400°C had the maximum mechanical performance. The authors ascertain that there is a good relationship between the microstructural change and the thermal characteristics of the developed coatings as microstructural buildup improved the thermal strengthening properties and was maintained.

According to Rahman et al. (2009), in order to strengthen the industrial application of mild steel for maximum usage, composite particulate is known as a functional material to enhance mild steel structural characteristics and relieve stress evolution. Cracks and stress evolve often seen on mild steel due to homogeneity and contamination. The proffers solution to this challenge deposited zinc–nickel alloys embedded in mild steel under various deposition conditions were used to enhance their structural properties. The effect of surface treatment and plating variables (bath composition, pH, and current density) on the coating composition, morphology, corrosion property, and microhardness was investigated. The result shows a modified morphology with perfect crystal growth, uniform arrangement of crystals, and refinement in crystal size, hence bright deposit was obtained from sulphate Bath-3 containing 30 g/L H_3BO_3 at a current density of 75 mA/cm^2. Corrosion resistance, as well as microhardness of Zn–Ni alloy coatings, increased with a %Ni increase in the deposit for all the sulphate baths studied. It was found that the characteristics of the deposited coating depend on the initial surface preparation, bath composition, additives, and temperature. The authors concluded that phases and microstructure of the surface of the deposited Zn–Ni alloys is another important characteristic which controls the corrosion resistance and other mechanical properties. One serious drawback during the electrodeposition process was the generation of internal stresses resulting in the formation of microcracks. However, the internal stress levels were reduced by buildup structure and heat treatment.

According to Rusu et al. (2012), one of the unique ways to enhance the microhardness behaviour and spread out the engineering application of mild steel is to develop a harder resistance coating. The authors study vividly the microstructural impact of the microhardness by varying the deposition parameters to induce changes in the structure and properties of the deposits. It was found that the microstructure of deposits changed with the current density and deposition time. The authors prepared and characterized thin Ni layers by electrodeposition through galvanostatic technique from a Watts bath at different current densities in the range from 1 to 10 A dm^2 and for deposition times between 900 and 7200 s. The structure and morphology of the nickel coatings were investigated by SEM and XRD techniques. The authors mentioned that a systematic experimental study was made to determine the role of the working parameters (current density and deposition time) on the structure and hardness properties. At a current density of 1 or 2 A dm^2, the nickel crystals were larger and better separated than at higher current densities. The uniform deposits obtained at 5 A dm^2 showed fine grains and better protection against corrosion. Microhardness was lowest for the layers electrodeposited at 10 A dm^2. The deposit hardness for the nickel Watts bath was highest for the layers electrodeposited at 1 A dm^2 and diminished when the current density was increased. In all, the author asserted that the excellent microstructural properties produced on the steel through the reinforced composite gave improved hardness properties observed.

5.4 ELECTRODEPOSITION OF ZINC-BASED PARTICLE ON MILD STEEL

The initial stages of Zn–Ni electrodeposition have been investigated by Lehmberg et al. (2005) through a study of the composition and structure of thin deposits on copper and mild steel substrates. Compositional gradients through the 0.05 μm thick deposits formed were characterized using glow discharge optical emission spectrometry (GDOES) and scanning transmission electron microscopy (STEM). The results obtained by GDOES show the influence of the substrate material on the composition of the deposits. Glancing-angle XRD (GAXRD) patterns confirm the formation of a complex phase structure in all the nickel–zinc deposits even though at the coating/substrate interface of the deposited alloy at 10 mA cm^{-2} the nickel content approached 60 wt.% and therefore one would expect a Zn–Ni phase to be formed. Ni–Zn deposits on a copper substrate have a relatively uniform composition.

Dikici et al. (2010: 787) investigated the mechanical and structural properties of Zn–Ni–Co ternary alloy electroplating on mild steel and contrasted it with the characteristics of Zn–Ni and Zn–Co alloy coatings. It was found that the obtained

Zn–Ni–Co alloy exhibited more preferred surface morphology and mechanical properties as compared to the other alloy coatings electroplated at the same conditions on mild steel. The microstructure of the composite by X-ray diffraction studies showed that the deposits of Zn–Ni–Co alloy coatings consisted of Zn, $ZnNi_3$, and $ZnCo_{13}$ phases. In addition, the hardness, elasticity modulus, and adhesion strength of coated alloys were measured with dynamic ultra-microhardness (DUH) and Scratch tester. The results of the Zn–Ni–Co alloy coating application have shown that the mechanical and surface characteristics of the coating layer are better than those of the Zn–Ni and Zn–Co alloy coating types in terms of displaying a homogeneous diffusion in this study. With ternary mild steel based, critical forces and adhesion strength of all the coatings are always visible and binary preferable to single coatings.

Fei et al. (2006) stated that "It is possible to electrodeposit zinc and Zn-Ni alloy CMM coatings using zinc sulfate electrolyte and the revised Zn-Ni alloy bath". The authors successively deposited zinc and Zn–Ni alloy compositionally modulated multi-layer (CMM) onto steel using dual baths. The coated samples were evaluated in terms of the surface appearance, surface and cross-sectional morphologies, as well as corrosion resistance. The microstructural characteristics that were examined using the field-emission gun scanning electron microscopy (FEGSEM) confirmed the layered structure and grain refinement of the zinc and Zn–Ni alloy CMM coatings and revealed the existence of microcracks caused by the internal stress in the thick Zn–Ni alloy sublayers. The corrosion resistance that was evaluated by means of the salt-spray test shows that the zinc and Zn–Ni alloy CMM coatings were more corrosion-resistant than the monolithic coatings of zinc or Zn–Ni alloy of the same thickness. The possible reasons for the better protective performance of Zn–Ni/Zn CMM coatings were given on the basis of the analysis of the micrographic features of zinc and Zn–Ni alloy CMM coatings after the corrosion test.

Hosking et al. (2007) explore the corrosion resistance performance of zinc–magnesium-coated steel in an acidified NaCl solution. The authors affirmed that zinc–magnesium-coated steel (ZMG) has greater corrosion resistance in a sodium chloride-containing atmosphere than conventional zinc-coated steel of similar coating thickness. A three-fold increase in time to red rust was recorded on ZMG versus conventional zinc-coated steel in an automotive cyclic corrosion test. The mechanism proposed for corrosion of ZMG in this work comprises the following steps: quenching of oxygen reduction reactions at cathodes by the formation of protective magnesium hydroxide. Decreased oxygen reduction facilitates the formation of stable oxide. Hydroxides associated with simonkolleite formation are neutralized by carbonization at the cathodes. Magnesium hydroxide at the cathodes is transformed into carbonate-containing hydroxides, which in turn facilitate further formation of protective simonkolleite.

Popoola and Fayomi (2011) studied the properties of zinc-plated mild steel in hydrochloric acid concentration in the presence and absence of an inhibitor. Zinc oxide was used in the investigation as retard inhibitor against acid variation through electrochemical polarization and weight loss technique. Experimental results revealed that the corroded layers of the zinc-plated mild steel without inhibitor have high content corrosivity related to the appearance of cracks due to the dissolution into the acidic medium. However, the zinc electro-deposition with ZnO as an inhibitor in the test medium shows a stable layer formation due to the insoluble complex ions. This helps to enhance the mild steel stability in a less aggressive acidic medium and mitigate the drastic corrosion attack that may occur. The Zn coatings were found to contain insoluble corrosion products, which act as a corrosion barrier but if the zinc-plated steel were exposed to an aggressive corrosion environment with acidic or oxide films, the deformation process would be activated within the surface boundary.

Pedroza et al. (2012) studied the effect of glycerol in Zn–Ni bath content to improve the corrosion resistance of carbon steel substrates in several applications, like aeronautical, oil and gas, and automotive industries. The deposition efficiency was analysed by galvanostatic and potentiodynamic deposition, and the corrosion resistance was estimated with electrochemical tests and mass loss measurements. The deposits were characterized by X-ray fluorescence, XRD, and SEM. The results of the investigation indicate that there is a significant increase in deposition efficiency and corrosion resistance when using glycerol concentrations up to 0.07 mol/L. For glycerol concentrations between 0.14 and 0.34 mol/L, there is a decrease in the deposition efficiency, and these concentrations do not result in a clear increase in the corrosion resistance of the Zn–Ni deposit. Analysis of the effect of deposition bath glycerol on the deposition efficiency and corrosion resistance of a Zn–Ni alloy indicates that this effect is related to the morphology of the deposit. The addition of glycerol to the deposition bath solution deposition results in grain refinement and the formation of a more compact deposit.

Chuen-Chang and Chi-Ming (2006) attempt to provide better corrosion protection by electrodeposition of zinc–nickel alloys on steel using pulse current. Some electric variables (average current density and pulse cycle) and bath temperature on chemical compositions and grain size of coatings were studied. In addition, the chemical compositions and the grain size of the Zn–Ni alloy coatings were determined by SEM combined with EDS and scanning probe atomic force microscope (AFM). From the author's observations, plating bath temperature was the main factor affecting the percentages of nickel in the deposits. However, the plating current densities and pulse cycle had little effect on the Ni compositions of deposits. Better deposition conditions for plating zinc–nickel in the deposits were assumed to be 313 K, 150 mA/cm^2, and Toff/Ton = 4, where grain size in the deposit reached a minimum. In addition, the rust layers analysed by FTIR were composed of $Zn(OH)_2$ and $ZnCl_2$, which confirmed the products of corrosion mechanisms. However, organic coatings applied over the Ni–Zn coating treatment would be considered for future work.

Thangaraj et al. (2009) carried out an investigation on composition-modulated alloy (CMA) electrodeposits of Zn–Co produced from acid chloride baths by the single-bath technique. Their corrosion behaviour was evaluated as a function of the switched cathode current densities and the number of layers. The process was optimized with respect to the highest corrosion resistance. Enhanced corrosion resistance was obtained when the outer layer was slightly richer with cobalt. From the author's observation at the optimum switched current densities of 40/55 mA/cm^{-2}, a coating with 600 layers showed six times higher corrosion resistance than monolithic Zn–Co electrodeposits having the same thickness. The CMA coating exhibited red rust only after 1130 hours in a salt-spray test. The increased corrosion resistance of the multilayer alloys was related to their inherent barrier properties, as revealed by electrochemical impedance spectroscopy. The corrosion rate of the CMA coating decreased as the number of layers increased, and the sequence of switched current densities was determined so that the outer, top layer of the coating had a higher concentration of cobalt. Even a small change in the content of cobalt in the layer was sufficient to change the corrosion resistance significantly. The electrochemical stability of the optimized CMA coating was explained in terms of an n-type semiconductor. It was demonstrated that optimization of the corrosion resistance is possible through proper manipulation of the deposition conditions and the structure of the coating.

Chitharanjan et al. (2010: 2031) research objectives aimed at comparative evaluation of Zn–Ni, Zn–Fe, and Zn–Ni–Fe coatings electrodeposited galvanostatically on mild steel from an acidic chloride bath containing gelatin and sulfanilic acid, with emphasis on their chemical composition, phase content, surface morphology, and corrosion resistance. According to the authors' knowledge, the bath compositions used in this study have not been reported before. The effect of bath additives is clarified by means of cyclic voltammetry. The comparison between the two binary alloys and the ternary alloy allows a better understanding of the role, individually and synergistically, of the two iron-group metals. Finally, the kinetics of alloy deposition is determined through calculations of the partial current densities. The result was used to elucidate the model of ACD which is relevant to these specific alloy systems. Cyclic voltammetry revealed that the deposition of the ternary alloy had distinguished characteristics compared to that of the binary alloys. For each current density, the concentration of either Ni or Fe in the ternary alloy was higher than its corresponding concentration in the binary alloys. This means that there could be a synergistic catalytic effect due to their co-presence in solution. SEM revealed that the surface morphology of the ternary alloy was significantly different from that of the binary alloys. AFM revealed that the ternary alloy had the lowest and most uniform surface roughness. XPS and XRD revealed the presence of ZnO in the bulk of the ternary alloy coating, not just on its surface. The ternary alloy exhibited the highest corrosion potential, lowest corrosion current density, highest electrochemical impedance, and highest slope of the Mott-Schottky line. Its enhanced corrosion performance is attributed to its significantly higher content of iron-group metals, the presence of ZnO both at the surface and in the bulk, smoother surface and improved protection by the surface film, which behaves like an *n*-type semiconductor.

Co-deposited Zn-submicron-sized Al_2O_3 composite coatings: Production, characterization, and micromechanical properties were investigated by Sancakoglu et al. (2011). The composite coatings were produced using different coating parameters such as current density, pH, temperature, agitation type, and ceramic powder content of the bath solution. It was found that co-deposition of submicron-sized Al_2O_3 particles and Zn metal was successfully achieved via the electrodeposition method without any chemical interaction between the ceramic particles and the electrolyte. A comparison of SEM images belonging to the coatings fabricated to those of the reference coatings revealed that a homogenous grain structure was obtained. Sakae and Daisuke (2007) studied the effect of zinc and zinc alloy-coated steel sheets on perforation corrosion in actual automobiles and the relevant accelerated corrosion test methods. The main factor affecting corrosion in the crevice of lapped panels was the coating weights of zinc and zinc alloys rather than the type of coating. In this study, the perforation corrosion resistance in zinc and zinc alloy-coated steel sheets was investigated quantitatively. Based on the analysis of actual automobiles and the evaluation of accelerated corrosion test methods, the corrosion resistance for perforation in an actual environment using accelerated corrosion was discussed. The Perforation Corrosion Index is considered the effective index for evaluation of the corrosion test conditions in various accelerated corrosion test methods in relation to the actual environment for automobiles.

5.5 DEPOSITION OF COMPOSITE-INDUCED METAL MATRIXES ON MILD STEEL

Gomes, Almeida et al. (2012) prepared Zn–Ni–TiO_2 and Zn–TiO_2 nanocomposites by galvanostatic cathodic square wave deposition. XRD analysis and SEM revealed that the inclusion of TiO_2 nanoparticles (spherical shaped with a diameter between 19.5 and 24.2 nm) promotes the formation of the Ni_5Zn_{21} phase, changes the preferred crystallographic orientation of Zn from (101) and (102) planes to (002), and decreases the particle size of the metallic matrixes. The stability of the nanocomposites immersed in near-neutral 0.05 m^3 mould Na_2SO_4 solution (pH 6.2) was investigated over 24 h. Data extracted from the steady-state polarization curves demonstrated that the metal–TiO_2 nanocomposites have, with respect to the metal coatings, a higher corrosion potential in the case of the Zn–Ni alloy composite, a lower corrosion potential in the case of Zn-based nanocomposite albeit the predominant (002) crystallographic orientation; and a lower initial corrosion resistance due to the smaller grain size and higher porosity in the Zn–Ni–TiO_2 and Zn–TiO_2 nanocomposites. Morphological and chemical analyses showed that a thicker passive layer is formed on the surface of the Zn–Ni–TiO_2 and Zn–TiO_2 deposits.

Praveen and Venkatesha (2008) carried out nano-sized TiO_2 particles prepared by sol–gel method. The TiO_2 particles were co-deposited with zinc from a sulphate bath at pH 4.5 using an electrodeposition technique. The corrosion behaviour of the coatings was assessed by electrochemical polarization, impedance; weight-loss and salt-spray tests. Wear resistance and microhardness of the composite coating were measured. The smaller grain size of the particulate was observed and it was confirmed by the images of SEM and XRD techniques. The Zn–TiO_2 composite coating generated by the electrolytic method indicated that nanoparticles were dispersed uniformly in the solution. The incorporation of TiO_2 in the coating led to improvement in the crystal size with smaller grain, thereby enhancing corrosion resistance, microhardness, and wear resistance properties of the composite coating as compared to the pure zinc coatings. According to the authors, the enhancement in the resistance is due to the physical barriers produced by TiO_2 to the corrosion process by filling crevices, gaps, and micron holes on the surface of the zinc coating. The TiO_2 uniformly distributed in the zinc coating and displaced the potential of the composite coating towards more positive values. The composite coating exhibits resilient corrosion properties. This excellent corrosion resistance, wear resistance, and microhardness of the composite coatings provide wide applications in modern industry.

TiO_2 nanoparticle coating has been applied on mild steel by Shanaghi et al. (2009), using sol–gel method. The coating was deposited on a mild steel substrate by hot-dip coating technique. The morphology and structure of the coating were analysed using SEM, AFM, and XRD. The anticorrosion performances of the coating were evaluated by electrochemical techniques. It is worth to note that the uniformity of films was retained in high temperatures and no crack and flake off from the substrate was observed. The Tafel polarization curves were employed to measure the anticorrosion performance of the TiO_2 coatings in a 3.5% NaCl solution. In conclusion, the authors affirmed that complete TiO_2 was uniformly embedded into the metal lattice without any crack with improved corrosion resistance.

Popoola et al. (2012) examined electrochemical and mechanical properties of mild steel electro-plated with Zn–Al particulate in an attempt to obtain a better surface adherent coating using electroplating technique. Zn–Al film was developed with zinc and aluminium powder particles dissolved in nitric acid and sodium hydroxide respectively, to form solutions containing Zn ions. ACD on mild steel resulted in surface modification attributed to the complex alloys that were developed. The microhardness value of the mild steel increased greatly due to the presence of Zn–Al particles from 55 HVN for the substrate to 137 HVN for the Zn–Al particulate. The microhardness of the coated sample clearly increased over two times as much as the base material. Wear resistance of the coated samples also increased and corrosion resistance of mild steel improved greatly.

The effect of SiC and Al_2O_3 nanoparticles in electrodeposited nickel coatings by Ortolani et al. (2012) was studied using XRD and SEM. The fabricated alloy entails pure nickel coating and two other coatings, respectively, with SiC and Al_2O_3 nanoparticles variation. The addition of Al_2O_3 and SiC shows good nanoparticle dispersion, as well as remarkable effects on Ni grain growth. The author further attested that the method is, therefore, a valid support not only for materials science studies which aim at optimizing coatings and thin films properties but also for a deeper understanding of underlying mechanisms connecting grain growth, texture formation, and the rise of residual stresses.

Ruidong et al. (2008) researched the effects of nano-CeO_2 concentrations in electrolytes on microstructures and properties of nanocomposite coatings. The samples were characterized for chemical compositions, element distributions, microhardness, and microstructures. Ni–W–P–CeO_2–SiO nanocomposite coatings were prepared on common carbon steel surface by pulse electrodeposition of nickel, tungsten, phosphorus, rare earth (nano-CeO_2), and silicon carbide (nano-SiO_2) particles. When nano-CeO_2 concentration in the electrolyte was controlled at 10 g/L, the electrodeposition sample particles led to uniform multiple-element nanocomposite coatings, possessing higher microhardness and compact microstructures with a clear outline of spherical matrix metal crystallites, fine crystallite sizes, and uniform distribution of every element within the Ni–W–P matrix metal. Increasing nano-CeO_2 concentrations in electrolytes from 4 to 10 g/L led to refinement in grain structure and the improvement of microstructures. While over 14 g/L, the crystallite sizes began to increase again and the smoothing degree decreased.

5.6 DEPOSITION OF COMPOSITE PARTICULATE BY CONVENTIONAL TECHNIQUES

The development of ceramic particulate-reinforced aluminium metal matrix composites (MMCs) has been studied by Durai et al. (2008), and the corrosion behaviour properties of Al–Zn–Al_2O_3 MMCs with minor alloying additions of Cu and Mn were verified. The composites were prepared via reaction sintering of partially reacted oxide mixtures derived from a high-energy ball milling process. The influence of high-energy ball milling on the corrosion behaviour of composites in a 3.5 wt% NaCl solution has been investigated through electrochemical experiments. The results of the corrosion tests evaluated using the potentiodynamic method in the NaCl solution indicate that the corrosion of the investigated composite materials depends on the weight fraction of the reinforcing particles. The polarization curves show that the milling procedure improves the corrosion resistance of the composites in the passive condition. The crystallite size of the mechanically milled Al powders decreases with the increasing milling duration, while the lattice strain increases with the milling duration.

According to Diserens et al. (1998), thin films of Ti–Si–N were deposited by PVD on stainless steel (type 304) substrates with the intention of improving the wear and hardness resistance. The authors demonstrated that the co-deposition of silicon nitride and titanium nitride clearly improves the mechanical properties of the coatings in terms of their hardness and abrasion resistance. The nanocomposite coatings show nano-sized crystallites of TiN and silicon nitride embedded into the metal matrix. The hardness of the TiN/SiN coatings reaches 3500 HV.

Shtansky et al. (2010) stated that the desired properties of multi-component nanostructured films can be achieved with hard films based on carbides, borides, and nitrides of transition metals by alloying with metallic (Al, Cr, and Zr) or non-metallic (O, P, Si, Ca) elements. The authors developed films for enhanced thermal stability, corrosion, and oxidation resistance. The results obtained showed that the composite gave significant improvement in relation to Young's modulus, low wear and friction, high thermal stability, oxidation, and corrosion resistance. Cheng et al. (2008) investigated the effect of silicon carbide particles on co-deposition of Ni–SiC using cyclic voltammetry and chrono-amperometry deposited test system. The significant influence of SiC particles on the nucleation and growth of nickel deposition on copper matrix from acid sulphate solution was achieved. The observation with SEM confirms that SiC particles can be considered favourable sites for nickel nucleation. The electron-crystallization behaviours of the Ni–SiC co-deposition process offer favourable help and reference for composite coating applications and research in the future. The author further explained that SiC particles may also mask the matrix surface and obstruct mass transmission for nickel deposition, which leads to a mild descending peak current. However, the particle cannot alternate the nucleation mechanism of nickel deposition on a matrix surface but has an influence on the deposition process and can refine the microstructure morphology of the composite coating.

Mou et al. (2010) carried out electrodeposition of Mg with Zn in acidic sulphate solutions with polyethylene glycol and octadecyldimethyl benzyl ammonium chloride as additives. Results obtained showed that these two compounds act in a synergetic way to suppress Zn deposition markedly and facilitate Mg reduction. This is most fundamental in zinc alloy deposition where additives may enhance or retard corrosion resistance. Based on this study the authors stated that during Zn coatings, hydrogen evolutions in neutral chloride solutions are visible compared to the produced Zn–Mg alloy coating. Magnesium hydroxide may cause current oscillations at high cathodic polarizations in plating solutions without zinc salts due to its formation and peel-off. An "induced co-deposition" mechanism is proposed for Zn–Mg alloy electrodeposition. From all indications, Mg^{2+} ions cannot be reduced in additive-free acidic sulphate solutions. Organic compounds polyethylene glycol (PEG) and octadecyldimethyl benzyl ammonium chloride (OC), used separately or together as additives, have a great effect on the cathodic polarization process. Compared with the single ones, a mixture of PEG and OC can shift zinc deposition potentials to more negative values. The synergetic effect between them creates a continuous barrier layer on the electrode surface and sufficiently suppresses zinc deposition under high current densities. As a result, Zn–0.46%Mg coatings are electrodeposited at −2 A cm^{-2} without burnt characteristics.

Basavanna and Arthoba Naik (2009) investigated the electrodeposition of Zn–Ni alloy from a chloride bath in the presence of condensation product (CP) formed between vanillin and hexamine. The nucleation growth and surface adhesion study was carried out on a glassy carbon electrode using cyclic voltammetry and chronoamperometric techniques. During the anodic scan of cyclic voltammetry, three anodic peaks were observed corresponding to the dissolution of zinc and nickel from different phases of Zn–Ni alloy. The model of Shivakumara et al. (2007) was used to analyse the current transients and it revealed that the Zn–Ni electrocrystallization process in the presence of CP, under the studied conditions, is governed by a three-dimensional nucleation process controlled by diffusion. In the presence of CP, the results indicated that the nucleation process changes from progressive to instantaneous when the deposition potential becomes more negative. The phase structure and surface morphology of the deposits were characterized by means of XRD analysis and SEM. With high deposition overpotential, the electrodeposition of Zn–Ni coating in the presence of CP followed the 3D nucleating and subsequent instantaneous grain growth mechanism, while in the case of low deposition over potential; it followed the 3D nucleating and subsequent progressive grain growth mechanism. SEM analysis of the coating morphologies showed that fine-grained deposits formed in the presence of CP, promoting the formation of smooth and shining coatings.

Gang et al. (2010) studied the composite coatings properties of Co–Ni–Al_2O_3 by using electrolytic co-deposition of Co–Ni alloys and Al_2O_3 from a sulphamate electrolyte containing Al_2O_3 particles. The influence of plating parameters on the content of the co-deposited Al_2O_3 particles in Co–Ni alloys was investigated. In the process of co-deposition, cathodic polarization increases with the increase of Al_2O_3 concentration in the bath, and cobalt ions in electrolyte cause the reduction of polarization while nickel ions do not change the polarization behaviour. Surface morphology and microstructure of Co–Ni–Al_2O_3 coatings determined by means of SEM, AFM, and XRD affirmed that the phase structure of solid solution cannot be varied by co-deposition of Al_2O_3 particles in Co–Ni alloys, but can only influence the growth and orientation of crystal planes. It was shown that the presence of Al_2O_3 particles in deposit greatly improves the hardness and the wear resistance of composite coatings. However, the co-deposition of Al_2O_3 increases the tensile internal stress of Co–Ni–Al_2O_3 deposit. The coefficient of thermal expansion and the thermal conductivity of Co–Ni–Al_2O_3 composite coatings are varied with the increase of Co contents and the temperature. In their investigation, surface morphology and microstructure of Co–Ni–Al_2O_3 coatings are mainly influenced by the cobalt contents. In the high-cobalt region, the coatings with hcp lattice structure have a more uniform and fine surface structure than those of the low-cobalt coatings with face-centred cubic (FCC) lattice structure.

Mohankumar et al. (2012) produced a thin film of Zn–Ni–Fe_2O_3 on steel substrates by electrodeposition technique using Zn–Ni alloy plating solution with nano-sized Fe_2O_3 particles. The cathodic polarization and cyclic voltammetry techniques were used to explain the deposition process. The corrosion behaviour of deposits was evaluated by polarization and impedance studies. In their investigation, Fe_2O_3 nanoparticles were dispersed in a Zn–Ni plating solution and a Zn–Ni–Fe_2O_3 coating was deposited on a steel substrate. Electrochemical studies were used on these composite coatings to establish their corrosion resistance properties. The result showed that the presence of Fe_2O_3 in the plating bath solution of the Zn–Ni alloy gave a Zn–Ni–Fe_2O_3 composite of smaller grain size. The composite was bright, compact, and possessed good corrosion resistance. The preferred orientation of Zn–Ni crystal changed in the presence of Fe_2O_3 during electrodeposition. All corrosion studies agreed that the composite gave better corrosion resistance and revealed that Fe_2O_3 particles held firmly to the Zn–Ni atoms in the alloy matrix, which made the dissolution difficult, hence Fe_2O_3 particles exhibited a reinforcement effect on the Zn–Ni alloy matrix.

Conde et al. (2011) electrodeposited Zn–Ni coatings on high-strength steel (4340) with a Ni content of about 14 wt% of pH = 13.6, which uses DETA as a complexing agent from an alkaline bath. A complete characterization of the coatings (corrosion, morphology, and composition) has been accomplished, correlating the electrodeposition conditions with these features. The best protective properties of the grown coatings were achieved for the alloys with a single-phase structure of c-Ni_5Zn_{21} and a denser morphology. Additionally, the hydrogen content incorporated is lower than even cadmium-coated 4340 steel which has undergone a post-baking dehydrogenation treatment. The authors elucidate their fact by saying the concentration of Ni in the coating depends on the stirring conditions. The structure and morphology of the layer varied with the applied current density. With a lower current density, the deposition of a layer with a practically single-phase structure was achieved, made up of the c-Ni_5Zn_{21} phase. Zn hexagonal phase appears together with the c phase at higher current densities. In addition, the lower the current density, the more compact the morphology of the coating. The best protective properties were achieved for the coatings with a single-phase structure and a denser morphology, these being the characteristics that determine the sacrificial properties of coatings and the long-term protective properties.

Murside and Alper (2011) investigated the effect of electrolyte pH and Cu concentration on the microstructure of electrodeposited Ni–Cu alloy films. Single Ni and Ni–Cu alloy films were electrodeposited on polycrystalline Ti substrates from sulphate electrolytes with different pH values under potentiostatic control. In Ni–Cu alloy films, it was found that the electrolyte pH does not affect the deposition processes, while the Cu concentration in the electrolyte has a considerable effect. The texture formation of Ni–Cu alloy films is not affected by the Cu concentration of the electrolyte. The EDX data showed that the Cu content of Ni–Cu alloy films increases with both increasing electrolyte pH and Cu concentration in the electrolyte. The SEM and AFM images showed that Ni films have smoother and more uniform surface than Ni–Cu alloy films. The surface morphology of Ni films is not affected by the electrolyte pH, while that of Ni–Cu alloy films is significantly affected by both the electrolyte pH and the Cu content. It was found that as the electrolyte pH decreases the Cu content and surface roughness of the films increases.

REFERENCES

Abdullah, M., Fouda, A. S., Shama, S. A., & Afifi, E. A. (2008). Azodyes as corrosion inhibitors for dissolution of C-steel in hydrochloric acid solution, Africa. *Journal of Pure and Applied Chemistry*, *9*, 83–91.

Arici, M., Nazir, H., & Aksu, A. (2011). Investigation of Sn-Zn electrodeposition from acidic bath on EQCM. *Journal of Alloys and Compound*, *509*, 1534–1537.

Basavanna, S., & Arthoba Naik, Y. (2009). Electrochemical studies of Zn–Ni alloy coatings from acid chloride bath. *Journal of Applied Electrochemistry*, *39*, 1975–1982.

Cheng-Yu, T., Yu, L., Xu-Shan, Z., & Zi-Qiao, Z. (2008). Nickel co-deposition with SiC particles at initial stage. *Transactions Non Ferrous Metals Society of China*, *18*, 1128–1133.

Chitharanjan, H. A., Venkatakrishna, K., & Eliaz, N. (2010). Electrodeposition of Zn–Ni, Zn–Fe and Zn–Ni–Fe alloys. *Surface and Coatings Technology*, *205*, 2031–2041.

Chuen-Chang, L., & Chi-Ming, H. (2006). Zinc-nickel alloy coatings electrodeposition by pulse current and their corrosion behavior. *Journal of Coating Technology and Research*, *3*(2), 99–104.

Conde, A., Arenas, M. A., & Damborenea, J. J. (2011). Electrodeposition of Zn-Ni coatings as Cd replacement for corrosion protection of high strength steel. *Corrosion Science*, *53*, 1489–1497.

Derek, A. O., Fayomi, O. S. I., & Atiba, J. O. (2024). Microstructural characterization, mechanical performance, and anti-corrosive response of zinc multifaceted coating on mild steel, *Key Engineering Materials*, *981*, 3–14.

Dikici, T., Culha, O., & Toparli, M. (2010). Study of the mechanical and structural properties of Zn–Ni–Co ternary alloy electroplating. *Journal of Coating Technology and Research*, *7*(6), 787–792.

Diserens, M., Patscheider J., & Lévy, F. (1998). Improving the properties of titanium nitride by incorporation of silicon. *Surface and Coatings Technology*, *108–109*, 241–246.

Dong, D., Chen, X. H., Xiao, W. T., Yang, G. B., & Zhang, P. Y. (2009). Preparation and properties of electroless Ni-P-SiO_2 composite coatings. *Applied Surface Science*, *255*, 7051–7055.

Durai, T. G., Das, K., & Das, S. (2008). Corrosion behaviour of Al-Zn/Al_2O_3 and Al-Zn-X/Al_2O_3 (X=Cu, Mn) composites synthesized by mechanical-thermal treatment. *Journal of Alloy and Compounds*, *462*, 410–415.

Ehteram, A., & Aish, H. (2008). Corrosion behavior of mild steel in hydrochloric acid solutions. *International Journal of Electrochemical Science*, *3*, 806–818.

Fayomi, O. S. I., Popoola, A. P. I., & Aigbodion, V. S. (2014). Effect of thermal treatment on the interfacial reaction, microstructural and mechanical properties of Zn–Al–SnO_2/TiO_2 functional coating alloys. *Journal of Alloys and Compounds, 617*, 455–463.

Fei, J. Y., Liang, G. Z., Xin, W. L., & Wang, W. K. (2006). Surface modification with zinc and Zn–Ni alloy compositionally modulated multilayered coatings. *Journal of Iron Steel Research International, 13*(4), 61–67.

Finazzi, G. A., Oliveira, E. M., & Carlos, I. A. (2004). Development of a sorbitol alkaline Cu–Sn plating bath and chemical, physical and morphological characterization of Cu–Sn films. *Surface and Coatings Technology, 187*, 377–387.

Fingsgar, M., & Jackson, J. (2014). Application of corrosion inhibitors for steels in acidic media for the oil and gas industry: A review. *Corrosion Science, 86*, 17–41.

Gang, W., Ning, L., Derui, Z., & Kurachi, M. (2010). Electrodeposited Co–Ni–Al_2O_3 composite coatings. *Surface and Coatings Technology, 176*, 157–164.

Gomes, A., Almeida, I., Frade, T., & Tavares, A. C. (2012). Stability of Zn–Ni–TiO_2 and Zn–TiO_2 nanocomposite coatings in near-neutral sulphate solutions. *Journal of Nanoparticles Research, 14*, 692–702.

Gomes, A., Frade, T., & Nogueira, I. D. (2012). Morphological characterization of Zn-based nanostructured thin films. *Current Microscopy Contributions to Advances in Science and Technology, 2*, 1146–1153.

Guo, S., Xu, L., Zhang, L., Chang, W., & Lu, M. (2012). Corrosion of alloy steels containing 2% chromium in CO_2 environments. *Journal of Corrosion Science, 63*, 246–258.

Han, W., Pan, C., Wang, Z., & Yu, G. (2014). A study on the initial corrosion behavior of carbon steel exposed to outdoor wet-dry cyclic condition. *Corrosion Science, 88*, 89–100.

Hosking, N. C., Strom, M. A., Shipway, P. H., & Rudd, C. D. (2007). Corrosion resistance of zinc–magnesium coated steel. *Journal of Corrosion Science, 49*, 3669–3695.

Hu, C., & Wang, C. (2006). Effects of composition and reflowing on the corrosion behavior of Sn-Zn deposits in brine media. *Electrochemical Acta, 51*, 4125–4134.

Jiang, L., Volovitch, P., Wolpers, M., & Ogle, K. (2012). Activation and inhibition of Zn-Al and Zn-Al-Mg coatings on steel by nitrate in phosphoric acid solution. *Journal of Science Corrosion, 60*, 256–264.

Lehmberg, C. E., Lewis, D. B., & Marshall, G. W. (2005). Composition and structure of thin electrodeposited zinc-nickel coatings. *Journal of Surface Coating and Technology, 192*, 269–277.

Li, M. C., Xin, S. S., & Wu, M. Y. (2010). Electrodeposition behaviour of Mg with Zn from acidic sulphate solutions. *Journal of Solid State Electrochemistry, 14*, 2235–2240.

Liu, Z., Wang, W., Wang, Y., Zhang, P., Wang, H., & Gao, C. (2014). Study of corrosion behavior of carbon steel under seawater film using the wire beam electrode method. *Corrosion Science, 80*, 523–527.

Mahieu, J., De Wit, K., De Boeck, A., & De Cooman, B. C. (1999). The properties of electrodeposited Zn-Co coatings. *Journal of Materials Engineering and Performance, 8*, 561–570.

Majumdar, J. D. (2010). Development of in-situ composite surface on mild steel by laser surface alloying with silicon and its remelting. *Journal of Surface and Coatings Technology, 205*, 1820–1825.

Melchers, R. E., & Chernov, B. B. (2010). Corrosion loss of mild steel in high temperature hard freshwater. *Journal of Corrosion Science, 52*, 449–454.

Mohankumar, C., Praveen, K., Venkatesha, V., Vathsala, K., & Nayana, O. (2012). Electrodeposition and corrosion behavior of Zn–Ni and Zn–Ni–Fe_2O_3 coatings. *Journal of Coating Technology Research, 9*(1), 71–77.

Montemor, M. F. (2004). Functional and smart coating for corrosion protection: A review of different advances. *Surface and Coating Technology*, *258*, 17–37.

Mou, C., Sen Sen, X., & Ming, Y. (2010). Electrodeposition behavior of Mg with Zn from acidic sulfate solutions. *Journal of Solid State Electrochemical*, *14*, 2235–2240.

Murside, H., & Mubel, A. (2011). Effect of electrolyte Ph and Cu concentration on microstructure of electrodeposited Ni-Cu alloy films. *Surface and Coating Technology*, *206*, 1430–1438.

Olivares-Xometl, O., Likhanova, N. V., Nava, N., Prieto, A. C., Lijanova, I. V., Escobedo-Morales, A., & López-Aguilar, C. (2013). Thiadiazoles as corrosion inhibitors for carbon steel in H_2SO_4 solutions. *International Journal of Electrochemical Science*, *8*, 735–752.

Ortolani, M., Zanella, C., Azanza, C. I., & Scardi, P. (2012). Elastic grain interaction in electrodepositional nano composite nickel matrix coatings. *Surface and Coating Technology*, *206*, 2499–2505.

Panagopoulos, C. N., Georgiou, E. P., Tsoutsouva, M. G., & Krompa, M. (2011). Composite multilayered coatings on mild steel. *Journal of Coating Technology Research*, *8*, 125–133.

Panossian, Z., De Almeida, N. L., De Sousa, R. M. F., De Souza, P. G., & Marques, L. B. S. (2012). Corrosion of carbon steel pipes and tanks by concentrated sulfuric acid. *A Review Corrosion Science*, *58*, 1–11.

Pedroza, G. A. G., Souza, C. A. C., Carlos, I. A., & Andrade Lima, L. R. P. (2012). Evaluation of the effect of deposition bath glycerol content on zinc–nickel electrodeposits on carbon steel. *Surface and Coatings Technology*, *206*, 2927–2932.

Popoola, A. P. I., & Fayomi, O. S. I. (2011). Performance evaluation of zinc deposited mild steel in chloride medium. *International Journal of Electrochemical Science*, *6*, 3254–3263.

Popoola, A. P. I., Fayomi, O. S. I., & Popoola, O. M. (2012). Comparative study of microstructure, tribological and corrosion properties of plated Zn and Zn alloy coating. *International Journal of Electrochemical Science*, *7*, 4860–4870.

Praveen, B. M., & Venkatesha, T. V. (2008). Electrodeposition and properties of Zn-nanosized TiO_2 composite coatings. *Applied Surface Science*, *254*(8), 2418–2424.

Rahman, M. J., Sen, S. R., Moniruzzaman, M., & Shorowordi, K. M. (2009). Morphology and properties of electrodeposited Zn-Ni alloy coatings on mild steel. *Journal of Mechanical Engineering, Transaction of the Mechanical Engineering Division, The Institution of Engineers, Bangladesh*, *40*, 9–12.

Ranganatha, S., Venkatesha, T. V., Vathsala, K., & Kumar, P. M. K. (2012). Electrochemical studies on Zn/nano-CeO_2 electrodeposited composite coatings. *Journal of Surface and Coatings Technology*, *208*, 64–72.

Roy, P., Karfa, P., Adhikari, U., & Sukul, D. (2014). Corrosion inhibition of mild steel in acidic medium by polyacrylamide grafted guar gum with various grafting percentage: Effect of intramolecular synergism. *Corrosion Science*, *88*, 246–253.

Ruidong, X., Junli, W., Zhongcheng, G., & Hua, W. (2008). Effect of rare earth non micro structures and properties of Ni-W-P-ClO_2-SiO_2 nano composite coating. *Journal of Rare Earth*, *26*, 579–584.

Rusu, D. E., Ispas, A., Bund, A., Gheorghies, C., & Cârâc, G. (2012). Corrosion tests of nickel coatings prepared from a Watts-type bath. *Journal of Coating Technology Resources*, *9*(1), 87–95.

Sakae, F., & Daisuke, M. (2007). Corrosion and corrosion test methods of zinc coated steel sheets on automobiles. *Corrosion Science*, *49*, 211–219.

Sancakoglu, O., Culha, O., Toparli, M., Agaday, B., & Celik, E. (2011). Co-deposited Zn-submicron sized Al_2O_3 composite coatings: Production, characterization and micromechanical properties. *Journal of Material and Design*, *32*, 4054–4061.

Shanaghi, A., Sabour, A. R., Shahrabi, T., & Aliofkhazraee, M. (2009). Corrosion protection of mild steel by applying TiO_2 nanoparticle coating via sol-gel method. *Protection of Metals and Physical Chemistry of Surfaces*, *45*, 305–311.

Shivakumara, S., Manohar, U., Arthoba Naik, Y., & Venkatesha, T. U. (2007). Influence of additives on electrodeposition of bright Zn–Ni alloy on mild steel from acid sulphate bath. *Bulletin Material Science*, *30*, 455–462.

Shtansky, D. V., Kiryukhantsev-Korneev, P. H. V., Bashkova, I. A., Sheveiko, A. N., & Levashov E. A. (2010). Multicomponent nanostructured films for various tribological applications. *International Journal of Refractory Metals and Hard Materials*, *28*, 32–39.

Soares, C. G., Garbatov, Y., Zayed, A., & Wang, G. (2009). Influence of environmental factors on corrosion of ship structures in marine atmosphere. *Corrosion Science*, *51*, 2014–2026.

Thangaraj, V., Eliaz, N., & Chitharanjan, A. (2009). Corrosion behavior of composition modulated multilayer Zn–Co electrodeposits produced using a single-bath technique. *Journal of Applied Electrochemistry*, *39*, 339–345.

Tiwari, S. K., Sahu, R. K., Pramanick, A. K., & Singh, R. (2011). Development of conversion coating on mild steel prior to sol gel nanostructured Al_2O_3 coating for enhancement of corrosion resistance. *Surface and Coatings Technology*, *205*, 4960–4967.

Tuaweri, T. J., & Wilcox, G. D. (2006). Behavior of Zn-SiO_2 electro deposition in the presence of N, N-dimethyldodecylamine. *Surface and Coating Technology*, *200*, 5921–5930.

Turk, A., Durman, M., & Kayali, E. S. (2007). The effect of manganese on the microstructure and mechanical properties of zinc-aluminium based Za-8 alloy. *Journal of Materials Science*, *42*, 8298–8305.

Xia, X., Zhitomirsky, I., & McDermid, J. R. (2009). Electrodeposition of zinc and composite zinc–yttria stabilized zirconia coatings. *Journal of Materials Processing Technology*, *209*(5), 2632–2640.

6 Corrosion Multifaceted Impact on Mild Steel

6.1 INTRODUCTION

The severe consequences of corrosion degradation have become a problem of worldwide significance. Corrosion is an electrochemical occurrence; it is the destructive mortification of a metal and alloy by chemical or electrochemical activities in its environment (Lin et al., 2012). Corrosion menaces cause plant shutdowns, waste of valuable resources, and reduction in efficiency and costly maintenance. It also jeopardizes safety and inhibits technological progress (Fuente et al., 2011). More so, the demand for fossil fuels has grown speedily, which contributes to increasing corrosion challenges with potential disasters that can cause serious issues, including negative social impacts, water resources, and environmental pollution. Corrosion problems in the oil field sector also occur at all stages from downhole to surface equipment and processing facilities leading to recurrent partial and eventual total process shutdown, resulting in severe economic losses. The other large problem in operating pipe flow lines is internal corrosion mainly due to stress corrosion cracking (Fingsgar & Jackson, 2014). This material destruction results in the loss of mechanical properties, such as strength, ductility, and impact strength. This results in loss of mechanical properties and at times ultimate failure. However, where corrosion damage is anticipated, the cost of maintenance and repair cannot be ignored since the costs vary from one industry and nation to the other. This could be either by direct or indirect cost. For instance, corrosion of metals costs in the U.S.A. economy is almost $300 billion per year at current prices. Approximately one-third of these costs could be reduced by the broader application of corrosion-resistant materials and the application of best corrosion-related technical practices (Li et al., 2014). The original work, based on an elaborate model of more than 130 economic sectors, found that metallic corrosion cost in the United States is about 4.9% of its gross national product (GNP), while in South Africa, it was also found that the cost was 4% GNP influence.

6.2 CORROSION CONTROL STRATEGIES AND ENVIRONMENTAL FACTORS

Corrosion control is achieved by recognizing and understanding corrosion mechanisms, by using corrosion-resistant materials and designs, and by using protective systems,

DOI: 10.1201/9781003562078-6

devices, and treatments. The effects of corrosion failures on the performance and maintenance of materials would often be minimized if life monitoring and control of the environmental and human factors supplemented efficient designs. One of the key factors in any corrosion situation is the environment (Tiwari et al., 2011; Guo et al., 2012). Variables such as time, humidity, and temperature instigate corrosion occurrences in major environments. It is also important to realize that the environment that affects a metal corresponds to the microenvironmental conditions that this metal really "sees", that is the local environment at the surface of the metal. It is indeed the reactivity of this local environment that determines the real corrosion damage. Most time, corrosion occurs as a result of atmospheric conditions (Soares et al., 2009). Atmospheric corrosion can be defined as the corrosion of materials exposed to air rather than immersed in a liquid. Atmospheric corrosion can be further classified into dry, damp, and wet categories. The effect of temperature on atmospheric corrosion rates is also quite complex. An increase in temperature will tend to stimulate corrosive attack by increasing the rate of electrochemical reactions and diffusion processes. For a constant humidity level, an increase in temperature would lead to a higher corrosion rate (Fingsgar & Jackson, 2014). The reactivity of this extreme corrosion propagation in relation to their environmental factors is also linked to the kind of metal or alloy involved (Guo et al., 2012; Melchers & Chernov, 2010). Basically, the impure and heterogeneous nature of some metals contributes immensely to the major cause of their poor corrosion resistance. The existence of anodic and cathodic sites on the metal surface and their reactivity with oxygen and water accelerate the change of a metal atom to a metal ion by the loss of electrons, which is also known as electrochemical corrosion (Popoola et al., 2012).

6.3 CORROSION VULNERABILITY OF MILD STEEL

Steel is nowadays a wide-ranging term covering an extensive area of iron alloys. Mild steel is one of the world's inexpensive and beneficial metals. Mild steel is simply a class of steel with very low hardness invariably with low ultimate tensile strength and ductility (Ranganatha et al., 2012). Its great demand to suit the requirements of different applications is incomparable to other metals. Certainly, mild steel has brought into being diverse uses in several fields, beginning from building construction to kitchen utensils, aerospace, and automobile (Majumdar et al., 2010). Mild steel is iron of crystalline alloy, carbon, and some additional elements, which hardens above its dire temperature to give a desired property. The manufacturing process of mild steel is relatively extensive as it comprises several stages of metallurgical decontamination (Panossian et al., 2012). The properties of mild steel hinge primarily on the extent of carbon it contains. Mild steel is a type of steel having a maximum carbon content of up to 0.25% along with small percentages of silica, sulphur, phosphorus, and manganese. Mild steel is very strong due to the low amount of carbon it contains. It has a high resistance to breakage. Mild steel, as opposed to higher carbon steel, is quite malleable, even when cold. This means it has high tensile and impact strength (Jiang et al., 2012). It is especially desirable for construction due to its good weldability and machinability. Mild steel is made into a wide range of products, including structural beams, car bodies, kitchen appliances, and cans. Although mild steel does have some appreciable properties, its limitations make it an engineering concern; these are as follows:

- Poor strengthening characteristics beyond 100,000 psi without major damage;
- Poor shape distortion and cracking of heat-treated steels;
- Poor impact resistance at low temperatures;
- Susceptibility to corrosion resistance for engineering challenges;
- Mild steel oxidizes freely at elevated temperatures (Soares et al., 2009).

Mild steel will corrode in many media, especially in most atmospheres. Generally, they are often selected not for their corrosion resistance performance but for economical and mechanical resilient behaviour such as strength, ease of fabrication, and cost. Mild steel can be susceptible to stress corrosion cracking in media that contain nitrates, hydroxides, ammonia, and hydrogen sulphide. Any evolved hydrogen may cause embrittlement and blistering in the steel (Popoola et al., 2012). Conceivably, most corrosion occurrence at major industrial outlets, such as electrical power plants or chemical processing facilities results in severe consequences which may cause the following:

- Changes of corroded component;
- Corrosion through excessive design;
- Protective preservation, such as painting (Mou et al., 2010);
- Closure of corrosion facilities due to corrosion catastrophe;
- Contagion of a conveyed product;
- Hurt of efficiency as a result of a decrease in the heat-transfer rate in heat exchangers;
- Destruction of equipment nearby to that in which corrosion let down takes place (Nahle et al., 2013).

Often, reactions in the corrosion process occur if there is an appropriate electron acceptor to associate with the electron discharge by the iron atom. In reality, the process of corrosion and the rate at which corrosion occurs are dependent on factors, which include but are not limited to a concentration of oxygen, acidity/alkalinity of pH level, water temperature, and dissolved salts like chloride, sulphate, and sulphites. However, there are different types of corrosion, which range from pitting to bacterial corrosion; they pose a huge threat, particularly to the marine, petrochemical, and automobile industries (Shibli et al., 2010; Sancakoglu et al., 2011). Zhu et al. (2011) stated that from a purely technical standpoint, an obvious answer to corrosion problems would be to use more resistant materials. In many cases, this approach is an economical alternative to other corrosion control methods. Corrosion resistance is not the only property to be considered in making material selections, but it is of major importance in the chemical process industries. The choice of a material is the result of several compromises (Dong et al., 2009). For example, the technical appraisal of an alloy will generally be a compromise between corrosion resistance and some other properties such as strength and hardness. Therefore, there is a need for corrosion scientists to study corrosion mechanisms to improve the understanding of the causes of corrosion and the ways to prevent or at least minimize damages caused by corrosion (Wang et al., 2011). Different ways that have been proposed to reduce corrosion by engineers include but are not limited to cathodic protection on a large scale for buried pipelines, new and better paints, prescribe proper dosage of corrosion inhibitors, and correct fabricated coatings (Nahle et al., 2013). The corrosion scientist and engineer, in turn, develop better criteria for cathodic protection, outline the molecular structure of chemical compounds that behave best as inhibitors, synthesize corrosion-resistant alloys, and recommend heat treatment and compositional variations of alloys that will improve their performance. Both the scientific and engineering viewpoints supplement each other in the diagnosis of corrosion damage and in the prescription of remedies.

6.4 SURFACE ENGINEERING FOR CORROSION MITIGATION

Surface engineering is both an art and a science of surface improvement. It is the treatment of a metallic surface through an application of surface-active agents or metallic coatings by electrochemical or electrolytic means either for altering the appearance, providing a protective coating, giving special surface properties, or improving engineering or mechanical properties (Dong et al., 2012). Surface protections are divided broadly into two categories metallic or non-metallic (organic or inorganic). With either of the two, the intent is the same, that is, to isolate the underlying metal from the corrosive media. Protective coatings are perhaps the most extensively used system for corrosion control. They are used to provide long-term protection under a broad range of corrosive environments, ranging from atmospheric contact to the most demanding chemical processing conditions. Protective coatings themselves provide little or no structural strength, yet they protect other materials to preserve their strength and integrity (Murside et al., 2011). Isolation of structural activeness of protective coating also gives proactive resistance against environmental corrosiveness. However, surface coatings must provide a continuous barrier to a substrate, and any imperfection can become the focal point for degradation and corrosion of the substrate if not carefully fabricated. The quality of a coating depends on many factors besides the nature of the materials involved. Coating and metal finishing operations are intended to increase corrosion or abrasion resistance, alter appearance, serve as an improved base for the adhesion of other materials, enhance wear and frictional characteristics, add hardness, and improve electrical properties (Yang et al., 2012). Surface coating is believed to be an engineering solution for the enhancement of surfaces counter to wear, corrosion degradation, and other surface-linked occurrences. Acceptable coatings are generally characterized by good adhesion, substrate compatibility, and low porosity (Popoola et al., 2012).

Coating must also be compatible with the physical constraints of the substrate such as temperature instability. With organic coatings, the primary function of corrosion protection is to detach the metal from the corrosive environment. In addition to forming a barrier layer to stifle corrosion, the organic coating can contain corrosion inhibitors (Muralidhara et al., 2012). Many organic coating formulations exist, as do a variety of application processes to choose from for a given product or service condition. Inorganic coatings include porcelain enamels, chemical-setting silicate cement linings, glass coatings and linings, and other corrosion-resistant ceramics. Like organic coatings, inorganic coatings for corrosion applications serve as barrier coatings. Some ceramic coatings, such as carbides and silicide, are used for wear-resistant and heat-resistant applications, respectively (Sancakoglu et al., 2011). Corrosion inhibitors as chemical species are known to either aggravate corrosion occurrence or act as surface-active agents for reducing corrosion progress. Chromates, silicates, and organic amines are common inhibitors that enhance corrosion resistance performance. The mechanisms of inhibition can be quite complex. In the case of the organic amines, the inhibitor is adsorbed on anodic and cathodic sites and stifles the corrosion current. Other inhibitors specifically affect either the anodic or cathodic process. Still others promote the formation of protective films on the metal surface (Loto & Popoola, 2011). The use of inhibitors in coating relieves defects and controls the corrosion progression.

The application of rational design principles can also eliminate many corrosion problems and greatly reduce the time and cost associated with corrosion maintenance and repair. Corrosion often occurs in dead spaces or crevices where the corrosive medium becomes more corrosive. These areas can be eliminated or minimized in the design process (Hassan et al., 2011). Where stress-corrosion cracking is possible, the components can be designed to operate at stress levels below the threshold stress for cracking. Where corrosion damage is anticipated, design can provide for maximum interchangeability of critical components and standardization of components (Majumdar et al., 2010). To improve on functional and high-performance applications, the fabrication of alloy through electrolytic co-deposition (electrodeposition) is an appreciated surface strengthening technique that provides outstanding wear deformation resistance, improved hardness, suitable temperature stability, and better corrosion performance of metal (Xu et al., 2008). The main target in the selection of functional composite particulates for advanced applications is their significant constituent of solid grains, the preparation of this alloy material, the mechanism behind the enhanced particle, and the re-crystallization affinity (Gomes et al., 2012). However, for excellent application, especially in high-temperature performance, high surface evolution characteristics are required and unification of high-temperature alloy particles has been proven to give such reinforcement property. According to Fayomi et al. (2014), regrettably, binary coating alloys selected for thermo-mechanical systems through electrodeposition are prone to possess possible limitations for high-temperature applications, and wear and corrosion environments. The author further affirmed that the control of process parameters is a major consideration in MMC co-deposition. The wear, corrosion resistance, thermo-mechanical stability, and tribo-oxidation behaviour of binary alloy composite coatings have been reported to give more stability at low-temperature applications (Wang et al., 2011).

Detailed information on surface coatings and modifications is necessary knowledge for recent applications in materials manufacturing design. Materials that range from ceramics, composites, biomaterials, and metals are of great benefit for suitable surface enhancement properties and achieving interfacial multi-functional systems where necessary. Surface coatings give good performance effects in oxidizing and corrosive media and provide valuable protection in high-temperature/high-stress applications. For example, ongoing research has shown that coatings can increase gas turbine engine efficiency via the application of thermal barrier coatings to superalloy-based turbine blades (Srivastava et al., 2010). Coatings can also boost performance in biomedical implant growth. A wide range of metal oxides exhibits fascinating properties and multiple functionalities. These materials have become the mainstream of practical studies. More so, the physical characteristics of metal oxides in relation to microstructure and structure buildup have been intensified by researchers lately (Gomes et al., 2012). A complete choice of these new device concepts can be foreseen through the incorporation of these noble materials with other scientifically vital materials. Several additive manufacturing technologies for direct metal fabrication and rapid manufacturing have emerged over the past two decades and some of them have reached industrial acceptance, for example in series production of medical implants and aerospace components. This evolution has largely occurred because of advancements in materials and processing equipment (Yang et al., 2012). Focused developments in metals, ceramics, polymers, and biomaterials have rendered them fully functional while being capable of assembly by one or another of the additive manufacturing processes. Numerous examples of parts in defence, commercial, and medical use give testament to the viability of these unique material processes that are capable of economical one-offs and otherwise unobtainable geometric configurations. The desire to design and assemble unique materials with improved mechanical and environmental performance that will give longer service life, and lower cost are ever in demand. Research advancement through materials fabrication, consideration of material failure, and progressive testing routes are crucial criteria to overcome these challenges (Hassan & Abdel Hamid, 2011).

Solution-based processing methods offer many advantages for the fabrication of composite and ceramic materials, including a high level of control over stoichiometry, microstructure, and morphology development, as well as tremendous flexibility in terms of materials and end-use architectures. Active research and development efforts continue in the development of materials and coatings with novel or improved properties using processes that offer enhanced control, reliability, reproducibility, and/or reduced cost, energy use and environmental impact, and related fabrication approaches (Mou et al., 2010). Twaweri and Wilcox (2006) established that the benefits of electrodeposition, especially for zinc coating over another coating, are the uniformity of deposition for intricate and multifaceted shapes, decrease of waste in deposition technique, low contamination levels, continuous and high production rates, and reasonable startup capital investment needed among others. Zinc electrodeposition with other alloys produces materials that endure the act of acid-chloride ions, organic solvents, and other harsh temperature hassles (Satapathy et al., 2009). Zinc depositions for steel protection are widely used for corrosion preservation due to their economic importance and cost efficiency. The report has affirmed that its corrosion protection is inadequate, especially in a concentrated environment (Popoola et al., 2012). Struggles to obtain novel materials for improved/combined properties for oxidation and instability of zinc coatings in recent times have been efforts towards the incorporation of MMC (Jiang et al., 2012).

According to Popoola and Fayomi (2011), zinc electroplating in steel industries constitutes a greater part of corrosion protection because of their sacrificial nature. However, the life span of such coatings is limited due to the hostile nature of some environments, principally those covering industrial pollutants. In a way to restrain this difficulty, surfactants are often employed (Praveen & Venkatesha, 2011). The use of surfactants could permit an exact composition of deposits to be attained. Facts have also been established that suitable surfactant does not only improve the stability of suspension by increasing the wettability of the particles but also boost the electrostatic adsorption of suspended particles on the cathode surface by increasing their positive charge (Tuaweri & Wilcox, 2006). Surfactant and organic compounds are known as additives applied to enhanced zinc plating and it has been widely studied to achieve robust and solid zinc coatings for corrosion prevention. These additives affect deposition and crystal-building progressions. From the adsorption phenomena, in order for surfactant incorporation to be effective, there is a chemical combination between the substrate and the adsorbate where electrons are transferred. Paunovic and Mordechay (2006) assert that if the rates of the adsorption and desorption processes are high and of the same order of magnitude as that of the cathodic deposition process, no incorporation or entrapment of additives in the deposit will occur. However, if they are much smaller, additive molecules will be entrapped in the deposit via propagating steps. Change in adsorbate concentration in solution can also result in orientational changes in deposits (Arici et al., 2011).

Propagation of microsteps to form macrosteps often occurs with the help of surfactants in zinc deposition. The type of deposit obtained at constant current density may depend on the assessment of the surface coverage by an additive leading to surface roughness during deposition. Sometimes, most coatings are made with multiple solvents and the selection of solvents impacts flow properties, viscosity, drying speed, and deposition characteristics. One of the serious problems associated with coatings is the erroneous choice of solvent because it can cruelly upset the curing and adhesion characteristics of the final coating (Xia et al., 2009). Some surfactant in deposition also serves as coating binder which provides uniformity and coherence to the coating system. Not all binders are corrosion-resistant; hence few serve in the preparation of protective coatings. The binder's capacity to produce a dense and tight film is directly linked to its molecular size and complexity (Ohgai et al., 2013). Binders with smaller molecular weights often react in situ. However, there are also consequences associated with the use of surface-active materials, which, in inappropriate concentrations, can obstruct with incorporated particle through extreme adsorption on the cathode surface. On the other hand, free surfactant can also be integrated into electrodeposited interface and cause adverse modifications in the mechanical behaviour like high internal stress and brittleness (Tuaweri & Wilcox, 2006).

Generally, zinc application with appropriate surfactant could not still be compared to the incorporation of transition oxide metal properties due to the aggressive nature of some environments as earlier mentioned (Gomes et al., 2012). The electrodeposition of suspended inert particles into a growing metal matrix seems to offer a possible solution to this problem and attracted research interest due to their unique functional properties such as improved corrosion and wear resistance, lubricity, semiconductor properties, magnetic properties, and superconduction (Yang et al., 2012). The electro-co-deposition properties attained generally hinge on the nature of particulate inclusion. Most composite coatings contain micron-sized particles and are acceptably considered because of their increasing accessibility, coupled with the capacity to fabricate interface with superior hardness, improved corrosion, and wear resistance (Fayomi et al., 2014). However, it is stated that the properties of alloy deposits are superior to those of single metal electroplates (Pedro et al., 2007). In other words, alloy co-deposition provides properties not obtained by electrodeposition of single metals and their properties depend strongly on the nature of the strengthening particles. These materials considered good candidates for surface alloy properties are basically harder, denser, possess suitable corrosion resistance, are more defensive of the underlying substrate metal, are more wear-resistant, have excellent low-friction properties, and have superior magnetic behaviour (Thangaraj et al., 2009). From the assertion made by Rusu et al. (2012), the plating parameters such as current density, rate of agitation, temperature range, bath constituents' concentration, and pH determine the fraction in which two or more metals uniformly co-deposit to have good surface-modified coatings. To this end, a significant variation in any one parameter could require a substantial and rewarding alteration in another variable or combination of parameters to maintain a given design and fabricated composite coating required.

A study undertaken by Rahman et al. (2009) indicated that no single variable or parameter has a self-determining effect on deposit composition or physical characteristics; all parameters can be considered with respect to their overall influence on the electrodeposition process. An increase in current density tends to increase the proportion of the less noble metal in the alloy deposit. Xia et al. (2009) pointed out that the extent of such change may be expected and could be greater in the case of simple primary salt solutions than in complex primary salt solutions due to different anions forming complex ions. In cases where the metals are associated with different complexing ions, a significant change in current density can be accommodated with relatively little change in plate composition. Since some composite additives influence the physical properties of deposits, limiting current densities can be altered by using the appropriate addition agents. With the diffusion process, additive agents or particulates migrate to the cathode through an applied force and cover the metal matrix with absorbed ions. These characteristics are often found in steel protected with tin to form $FeSn_2$.

However, the most intriguing report by Rusu et al. (2012) identified that transition metal oxides exhibit fascinating properties and multiple functionalities; and with their sensitive nature provide improved microstructure, reinforced crystal grains, and better mechanical properties for engineering application. These surface modification characteristics of composite alloys have drawn keen commercial interest for a wide range of structural applications due to their high strength at elevated temperatures, good thermal conductivity, superior oxidation resistance, excellent wear resistance, and reduced density compared to superalloys and other high-temperature materials (Zhu et al., 2011). Invariably, composite materials are now gaining adoption in mainstream applications for enhancing the performance of components and reducing structural weight (Fayomi & Popoola, 2013). Titanium oxides have traditionally been used in aerospace, medical, automotive, and sporting goods applications due to their high strength and stiffness-to-weight ratio and excellent biocompatibility. On the other hand, their variety in niche applications is numerous. There is also continued interest in increasing the use of tin alloy by developing lower-cost processes, improved properties, and improved biocompatibility. Arici et al. (2011) reported on the novel effect of tin on Sn–Zn alloy as sacrificial coating protection of ferrum-based metals. The property of the coatings was affirmed to provide high corrosion protection, frictional and anti-frictional properties, ductility, and solderability to the base metal. Composite characteristics are influenced by the deposition parameters and these have a significant effect on the surface modification properties which have been reported by various authors. Electro-co-deposition route has been explored with the help of different composite additives in a single series for mild steel improvement in the past. The goal of this work is to develop a functional bath and fabricate an operational zinc-rich thin film alloy in the presence of aluminium, tin, and titanium composite particulates. However, the target is aimed at improving the corrosion, mechanical, and tribological properties of mild steel. This is because the roles of composites in engineering practices, especially coating applications in recent times are enormous.

REFERENCES

Arici, M., Nazir, H., & Aksu, A. (2011). Investigation of Sn-Zn electrodeposition from acidic bath on EQCM. *Journal of Alloys and Compound*, *509*, 1534–1537.

Dong, D., Chen, X. H., Xiao, W. T., Yang, G. B., & Zhang, P. Y. (2009). Preparation and properties of electroless Ni-P-SiO_2 composite coatings. *Applied Surface Science*, *255*, 7051–7055.

Dong, Z., Peng, X., Guan, Y., Li, L., & Wang, F. (2012). Optimization of composition and structure of electrodeposited Ni-Cr composites for increasing the oxidation resistance. *Corrosion Science*, *62*, 147–152.

Fayomi, O. S. I., & Popoola, A. P. I. (2013). Chemical interaction, interfacial effect and microstructural characterization of the induced zinc-aluminuim–solanium tuberosum in chloride solution on mild steel. *Research Chemical Intermediates*, *39*, 1354–1564. Doi.10.1007/S11164-013-1354-2

Fayomi, O. S. I., Popoola, A. P. I., & Aigbodion, V. S. (2014). Effect of thermal treatment on the interfacial reaction, microstructural and mechanical properties of Zn–Al–SnO_2/TiO_2 functional coating alloys. *Journal of Alloys and Compounds*, *617*, 455–463.

Fingsgar, M., & Jackson, J. (2014). Application of corrosion inhibitors for steels in acidic media for the oil and gas industry: A review. *Corrosion Science*, *86*, 17–41.

Fuente, D., Diaz, I., Simancas, J., Chico, B., Morcillo, M. (2011). Long term atmospheric corrosion of mild steel. *Corrosion Science*, *53*, 604–617.

Gomes, A., Frade, T., & Nogueira, I. D. (2012). Morphological characterization of Zn-based nanostructured thin films. *Current Microscopy Contributions to Advances in Science and Technology*, *2*, 1146–1153.

Guo, S., Xu, L., Zhang, L., Chang, W., & Lu, M. (2012). Corrosion of alloy steels containing 2% chromium in CO_2 environments. *Journal of Corrosion Science*, *63*, 246–258.

Hassan, H. B., & Hamid, Z. A. (2011). Electrodeposited Ni-Cr_2O_3 nanocomposite anodes for ethanol electro-oxidation. *International Journal of Hydrogen Energy*, *36*, 5117–5127.

Jiang, L., Volovitch, P., Wolpers, M., & Ogle, K. (2012). Activation and inhibition of Zn-Al and Zn-Al-Mg coatings on steel by nitrate in phosphoric acid solution. *Journal of Science Corrosion*, *60*, 256–264.

Li, J. K., Ma, H. X., Zhu, S. D., Qu, C. T., & Yin, Z. F. (2014). Erosion resistance of CO_2 corrosion scales formed on Api P110 carbon steel. *Corrosion Science*, *86*, 101–107.

Lin, Z., Li, X., & Xu, L. (2012). Electrodeposition and corrosion behavior of zinc-nickel films obtained from acid solutions: Effects of TEOS as additive. *International Journal of Electrochemical Science*, *7*, 12507–12517.

Loto, C. A., & Popoola, A. P. I. (2011). Effect of tobacco and kola tree extracts on the corrosion inhibition of mild steel in acid chloride. *International Journal of Electrochemical Science*, *6*, 3264–3276.

Majumdar, J. D. (2010). Development of in-situ composite surface on mild steel by laser surface alloying with silicon and its remelting. *Journal of Surface and Coatings Technology*, *205*, 1820–1825.

Melchers, R. E., & Chernov, B. B. (2010). Corrosion loss of mild steel in high temperature hard freshwater. *Journal of Corrosion Science*, *52*, 449–454.

Mou, C., Sen Sen, X., & Ming, Y. (2010). Electrodeposition behavior of Mg with Zn from acidic sulfate solutions. *Journal of Solid State Electrochemical*, *14*, 2235–2240.

Muralidhara, H. B., Naik, Y. A., Balasubramanyam, J., Kumar, K. Y., Hanumanthappa, H., & Veena, M. (2012). Nanocrystalline zinc coating on steel substrate using condensation product of glycyl-glycine (Ggl) and vanillin (Vnl) and its corrosion study. *International Journal of Chemical Sciences*, *10*(1), 524–538.

Murside, H., & Mubel, A. (2011). Effect of electrolyte Ph and Cu concentration on microstructure of electrodeposited Ni-Cu alloy films. *Surface and Coating Technology*, *206*, 1430–1438.

Nahle, A., Sl-Khayat, M., Abdoun, I., & Abdel-Rahman, I. (2013). Corrosion inhibition of mild steel by p.p-bis (triphenylphosphonio) methyl benzophenone dibromide in HCl solution. *Anti-Corrosion Methods and Material*, *60*, 20–27.

Ohgai, T., Ogushi, K., & Takao, K. (2013, March). Morphology control of Zn-SiO2 composite films electrodeposited from aqueous solution containing quaternary ammonium cations. In *Journal of Physics: Conference Series* (Vol. 417, No. 1, p. 012006). IOP Publishing.

Panossian, Z., De Almeida, N. L., De Sousa, R. M. F., De Souza, P. G., & Marques, L. B. S. (2012). Corrosion of carbon steel pipes and tanks by concentrated sulfuric acid. *A Review Corrosion Science*, *58*, 1–11.

Paunovic, M. & Mordechay, S. (2006). Fundamentals of Electrochemical Deposition. *Joh Wiley & Sons*.

Pedro De Lima, N., Adriana, N., Correia, P., & Walney, S. A. (2007). Corrosion study of electrodeposited Zn and Zn-Co coatings in chloride medium. *Journal of Brazil Chemical Society*, *18*, 1164–1175.

Popoola, A. P. I., & Fayomi, O. S. I. (2011). ZnO as corrosion inhibitor for dissolution of zinc electrodeposited mild steel in varying HCl concentration. *International Journal of the Physical Sciences*, *6*, 2447–2454.

Popoola, A. P. I., Fayomi, O. S. I., & Popoola, O. M. (2012). Electrochemical and mechanical properties of mild steel electro-plated with Zn-Al. *International Journal of Electrochemical Science*, *7*, 4898–4917.

Praveen, B. M., & Venkatesha, T. V. (2011). Electrodeposition and corrosion resistance properties of Zn-Ni/TiO_2 nano composite coating. *International Journal of Electrochemistry*, *261*, 407–410.

Rahman, M. J., Sen, S. R., Moniruzzaman, M., & Shorowordi, K. M. (2009). Morphology and properties of electrodeposited Zn-Ni alloy coatings on mild steel. *Journal of Mechanical Engineering, Transaction of the Mechanical Engineering Division, The Institution of Engineers, Bangladesh*, *40*, 9–12.

Ranganatha, S., Venkatesha, T. V., Vathsala, K., & Kumar, P. M. K. (2012). Electrochemical studies on Zn/nano-CeO_2 electrodeposited composite coatings. *Journal of Surface and Coatings Technology, 208*, 64–72.

Rusu, D. E., Ispas, A., Bund, A., Gheorghies, C., & Cârâc, G. (2012). Corrosion tests of nickel coatings prepared from a Watts-type bath. *Journal of Coating Technology Resources, 9*(1), 87–95.

Sancakoglu, O., Culha, O., Toparli, M., Agaday, B., & Celik, E. (2011). Co-deposited Zn-submicron sized Al_2O_3 composite coatings: Production, characterization and micromechanical properties. *Journal of Material and Design, 32*, 4054–4061.

Satapathy, A. K., Gunasekaran, G., Sahoo, S. C., Amit, K., & Rodrigues, P. V. (2009). Corrosion inhibition by justiciagendarussa plant extract in hydrochloric acid solution. *Journal of Corrosion Science, 51*, 2848–2850.

Shibli, S. M. A., Chacko, F., & Divya, C. (2010). Al_2O_3-ZrO_2 mixed oxide composite incorporated aluminium rich zinc coatings for high wear resistance. *Journal of Corrosion Science, 52*, 518–525.

Soares, C. G., Garbatov, Y., Zayed, A., & Wang, G. (2009). Influence of environmental factors on corrosion of ship structures in marine atmosphere. *Corrosion Science, 51*, 2014–2026.

Srivastava, M., Balaraju, J. N., Ravishankar, B., & Rajam, K. S. (2010). Improvement in the properties of nickel by nano-Cr_2O_3 incorporation. *Surface and Coatings Technology, 205*, 66–75.

Thangaraj, V., Eliaz, N., & Chitharanjan, A. (2009). Corrosion behavior of composition modulated multilayer Zn–Co electrodeposits produced using a single-bath technique. *Journal of Applied Electrochemistry, 39*, 339–345.

Tiwari, S. K., Sahu, R. K., Pramanick, A. K., & Singh, R. (2011). Development of conversion coating on mild steel prior to sol gel nanostructured Al_2O_3 coating for enhancement of corrosion resistance. *Surface and Coatings Technology, 205*, 4960–4967.

Tuaweri, T. J., & Wilcox, G. D. (2006). Behavior of Zn-SiO_2 electro deposition in the presence of N, N-dimethyldodecylamine. *Surface and Coating Technology, 200*, 5921–5930.

Wang, G., Zhu, L., Liu, H., & Li, W. (2011). Zinc-graphite composite coating for anti-fouling application. *Journal of Material Letters, 65*, 3095–3097.

Xia, X., Zhitomirsky, I., & Mcdermid, J. R. (2009). Electrodeposition of zinc and composite zinc–yttria stabilized zirconia coatings. *Journal of Materials Processing Technology, 209*, 2632–2640.

Xu, R., Wang, J., Guo, Z., & Wang, H. (2008). Effect of rare earth on microstructures and properties of Ni-W-P-CeO_2-SiO_2 nano-composite coating. *Journal of Rare Earth, 26*, 579–583.

Yang, G., Chai, S., Xiong, X., Zhang, S., Yu, L., & Zhang, P. (2012). Preparations and tribological properties of surface modified Cu nanoparticles. *Transactional Non Ferrous Metals Society of China, 22*, 366–372.

Zhu, X., Cai, C., Zheng, G., Zhang, Z., & Li, J. (2011). Electrodeposition and corrosion behaviour of nanostructured Ni-tin composite films. *Transactional Non Ferrous Metals Society of China, 21*, 2216–2224.

7 Aluminium Composite Development for Sustainable Production

7.1 INTRODUCTION

Aluminium and its alloys are widely employed in various industries and applications, including transportation, aerospace, construction, and packaging. Aluminium and its alloys have remarkable qualities, making them one of several metallurgical materials that are widely employed in industrial and technical applications, ranging from engine blocks to aluminium beverage cans, laminated films, and extremely ductile wrapping foil. The primary reason for its use in many technical applications is due to its low density, formability, high conductivity, and high heat conductivity. Apart from the benefits outlined above, however, the technical applications for aluminium and alloys are limited due to poor surface characteristics and low abrasion resistance (Ayogu & Eze, 2019). Cu, Ce, Cr, Fe, Mg, Ti, In, and Zr have all been employed to improve the characteristics of aluminium for purposes over the years. Aluminium has a density of 2.7 kg/cm^3 in its normal condition, which is about one-third that of steel. A cubic foot of steel weighs around 490 pounds, while a cubic foot of aluminium weighs approximately 170 pounds (Dhanesh et al., 2021). This low mass density, combined with the high strength of some aluminium alloys, which exceeds that of structural steel, allows for the design and construction of strong lightweight structures that are especially beneficial for anything that moves—spacecraft and aircraft, as well as all types of land and water-borne vehicles (Mansor et al., 2019). Because various cold working procedures are involved in realizing a desirable output, the mechanical performance of aluminium and its alloying element plays a vital part in the product of aluminium packaging industries.

A recycled aluminium feedstock of grade 3014 HO9 was alloyed using the stir casting process to improve the desired mechanical performance. Aluminium is a lightweight, silvery-white metal formed from bauxite ore that exists in association with oxygen as alumina. It is used to make cans, foils, and packaging made of laminated paper or plastic. Magnesium and manganese, as composites, play an important role in enhancing the strength and characteristics of aluminium (Maneiah et al., 2020). The tensile strength of 90 MPa for pure aluminium can be enhanced through heat treatment for some specified heat-treatable alloys. The effect of air, moisture, and chemicals on the surfaces of aluminium materials is reduced by covering them with oxide. Aluminium is an excellent recycling material since it is simple to reclaim and turn

DOI: 10.1201/9781003562078-7

into new products. Aluminium has zero waste in the packaging industry since practically every offcut or scrap generated at various stages has economic use. However, the numerous processes that aluminium alloys go through in the packaging industry are the foundation for evaluating aluminium alloys with alloying components that can survive cold working. The sustainability or impurities of alloying components are considered; also, impurity elements in some alloys may be important elements in others (Rana et al., 2012). Unlike many other metals, aluminium's high resistivity does not negate the need for an oxide layer to protect it from corrosion, air, temperature, moisture, and chemical attack. Aside from its exceptional barrier capabilities against moisture, air, light, and microorganisms, aluminium's mechanical properties include flexibility and remarkable surface resistance, excellent malleability and formability, and outstanding embossing capability. After steel, aluminium takes the lead in terms of metal and non-ferrous metal production volume, with more volume produced than all other non-ferrous metals combined.

Particularly in the aerospace and automotive industries, aluminium is becoming a favoured material over steel components. It is one of the most recycled materials on the planet due to its economic worth and widespread application in the construction, packaging, automotive, aerospace, and electrical distribution industries (Saikrupa et al., 2021). According to the Aluminium Association, 90% of aluminium used in construction and automotive parts is recycled at the end of its useful life. 75% of all aluminium produced in the past is still in use today (Doan et al., 2021). In 2010, the amount of primary and secondary aluminium produced in the United Kingdom was roughly 343 and 260 kt, respectively (McKenna & Norman, 2010). Aluminium's reactivity, particularly with oxygen, makes it extremely difficult to find in its elemental form, which justifies its extraction from minerals in ores. In 2016, the globe mined 289 million tons of bauxite ore, produced 118.9 million tons of aluminium oxide (alumina), and melted 58.8 million tons of primary aluminium (Daniel et al., 2020). The rise in these values is expected, in part, due to advances in aluminium alloy metallurgical methodology, population growth, and economic activity necessitating a bigger volume. The global demand for aluminium is expected to rise, according to a variety of forecasts. According to Yi et al. (2024), aluminium consumption will double in the next ten years, while others anticipate demand will double or triple by 2050.

Primary aluminium, on the other hand, has better mechanical qualities than secondary alloys. Most commercial alloys have desirable qualities that can be improved by adding alloys and employing heat treatment to alter heterogeneous microstructures. As material strength improves through a solid solution and hardening precipitation, the addition of alloying elements to matrix elements may impact the wear qualities of Al–Si–Mg. Fe, Cr, Cu, Mg, and Zn are alloying elements that improve mechanical characteristics (Kaya & Aker, 2017). Numerous studies have been conducted in recent years to determine the impact of specific intermetallic particles and individual alloying elements on the corrosion resistance of Al alloys as a result of pitting and intergranular-type corrosion. Within Al alloys, intermetallics have been demonstrated to be precise (Mg_2Si, $MgZr_2$, $Al_2Cu_2Mn_3$, Al_2Mn_3Si, Al_7Cu_2Fe, $AlCu_9$, Al_2CuMg, Al_3Fe, $Al_{12}Mg_2Cr$, Al_6Mn_9, $Al_2Cu_2Mn_{39}Al_3Ti$, Mg_2Al_3, Al_3Zr, and $Al_{32}Zn_{49}$). Aside from Mg_2Si and Al_3Ti, the alloying elements indicated above improve both mechanical and corrosion qualities. Furthermore, aluminium alloys are divided into three groups: constituent particles (Al_3Fe, Al_7Cu_2Fe, AlFeSi, and other AlFe particles), precipitates ($MgZn_2$, Mg_2Al_3, Mg_2Si, and Al_2Cu), and dispersed particles ($MgZn_2$, Mg_2Al_3, Mg_2Si, Al_2Cu) (Al_3Ti, A_6Mn, $Al_{20}Cu_2Mn_3$, and Al_3Zr). Precipitates are generated by nucleation and growth in a supersaturated solid solution during natural or low-temperature artificial ageing and range in size from Angstroms to fractions of a micrometre. With the inclusion of a common aluminium–silicon alloy, the exceptional features of aluminium, such as superior castability, corrosion resistance, good wear properties, low melting point and good fluidity, and good machinability, are brought to bear.

A higher silicon ratio in the elemental matrix material minimizes thermal expansion, improves wear characteristics, and decreases machinability (Bandil et al., 2019; Nallusamy, 2016). However, the mechanical characteristics of silicon are influenced by its composition and shape, regardless of the temperature range (level). The shape of the silicon phase is refined by chemical structural alteration, which changes the silicon from a plate-like to a fibrous structure (Haghayeghi et al., 2018). Impurity induces twinning and twin plane re-entrant edge growth are two well-established growth models that support these alterations. In addition, thermal alterations (quench modification) are another method for purifying silicon (Yan et al., 2019). MMC in a soft aluminium matrix depicted by the hard silicon phases. The use of silicon chains to improve load transfer is one component that contributes to increased material strength. Long-term high-temperature exposure of aluminium–silicon alloys produces coarsening of the eutectic silicon phases, resulting in a significant loss of mechanical properties (Zamani et al., 2017). By solid solution strengthening and dispersion strengthening methods, copper and magnesium are frequently employed to boost the material strength of alloys at room temperature and increased temperatures (Gazizov & Kaibysher, 2017). Copper has a maximum solid solubility of 5.65 wt% in aluminium at 545°C, whereas magnesium has a maximum solid solubility of 14.9 wt% at 451°C. Advantageously, researchers have demonstrated that using transition elements at elevated temperatures improves mechanical qualities over a long period of time. The high cost of rare earth elements, on the other hand, is a major impediment to their economic application.

When designing novel alloys for usage at extremely high temperatures, the economic advantages of transition elements nickel and cobalt over rare elements attract more attention (Reichardt et al., 2021). At high temperatures, nickel inclusion increases the strength and hardness of aluminium–silicon by forming thermally resistant Ni-rich compounds such as Al_3Ni, Al_9FeNi, Al_3CuNi, and others (Rakhmonov et al., 2016). The intermetallic compounds formed in the interdendritic areas during solidification give substantial strength at elevated temperatures. Nickel solid solubility in aluminium ranges from 0% to 0.04% by weight. The use of titanium as a master alloy refines the grain sizes of most aluminium alloys. In the melt, titanium is found as Al_3Ti dissolves and TiB_2 particles. However, the Al_3Ti dissolves in the melt, and the TiB_2 settles and agglomerates with time. During solidification, the TiB_2 particles exhibit stable phases and also act as heterogeneous nuclei, and the effect of grain refinement is over time (Xiao et al., 2019). Peritectic Al_3Ti develops in the melt when the solubility limit in the x—aluminium titanium is higher than in titanium. As Al_3Ti forms on the TiB_2 particles, the nucleation efficiency increases (Greer, 2016). Grain refining of aluminium silicon alloys through inoculation with master alloys such as Al-5Ti-1B1, Al-4B, and others improves feeding, castability, and shrinkage porosity (Riestra et al., 2017). Some metals' material impurities are fundamentally responsible for their poor mechanical performance and corrosion resistance. The chemical behaviour of metal anodic and cathodic areas, as well as their proclivity to react with water and oxygen, accelerates the pace of metal atom-to-ion transformation by electron loss (Zhao et al., 2020). Aluminium is now one of the world's most affordable and versatile metals. Aluminium's suitability for various purposes distinguishes it from other metals, and this is the reason for its great demand. Cookware, many sectors of technical projects, aircraft, and automobiles are only a few of the many areas of application (Fayomi & Akande, 2019). In terms of mechanical, chemical, and physical properties, aluminium is similar to brass, zinc, steel, copper, titanium, and lead. They are all metals that can be melted, cast, and machined in the same way. Apart from a few drawbacks, aluminium's remarkable characteristics make it a desirable material for engineering components and other household equipment. The following are the distinguishing characteristics:

- Natural production of protective thin films of coating on the surface of the substance from the environment.
- Compared to the steel of the sample dimension, it is lightweight.
- High heat and electrical conductivity, nearly twice that of copper.
- Good ductility, low density, and low melting point. Its non-magnetic properties make it extensively used for electrical shielding as in dish antennas and bus bars.
- Due to its excellent shock and sound absorber, it has found applications in auto bumpers and ceiling construction.
- Recyclability: 100% recyclable and virtually indistinguishable from the original product.
- Both primary and secondary production have low energy requirements.
- Due to its softness, it does not emit sparks when it comes into touch with itself or other materials. Temperature reduction favours its tensile strength, unlike steel which suddenly becomes brittle at low temperatures.
- Has a reasonable tensile strength and hardness but is weaker than stainless steel (Tisza & Czinege, 2018; Fayomi et al., 2019).

The addition of alloying elements such as magnesium, copper, and other elements, followed by heat treatments such as precipitation hardening or age hardening, are processes that can be used to strengthen metal alloys. Pure aluminium and aluminium alloyed primarily with magnesium or manganese are unaffected by heat treatment. Heat treatment is beneficial to aluminium alloys that contain zinc, copper, or a combination of silicon and magnesium as constituent alloying elements (Stemper et al., 2019). To improve the hardness of high-purity or non-alloyed metals, grain size reduction and strain hardening are frequently used. While heat treatment does not affect the qualities of all aluminium alloys, it does increase the ease of forming and the strength of the finished product of certain alloys. In comparison to steel, consistent heat management is essential for aluminium to achieve the best outcomes.

As a result, specialized equipment is usually necessary. All alloys benefit from cold working procedures such as wire drawing and cold rolling. However, varied combinations of cold working, heat treatment, and alloying can improve the strength of aluminium alloys (Gu et al., 2016). Because of their excessive softness and reactive nature, engineering components and utensils made of raw aluminium that are exposed to high temperatures are unsuitable (Vu et al., 2019). As a result, anodizing such a component protects it from thermal pitting. The existing aluminium cookware is anodized to protect it from the effects of high temperatures.

This is accomplished by treating the wares with chemicals to harden and minimize their reactivity. The purpose of anodizing aluminium is to enhance the thickness of the natural oxide layer of the surface by using an electrolytic passivation technique or anodic films. By anodizing, the item to be treated becomes the anode of the electrolytic cell, increasing wear resistance, corrosion resistance, and paint primer adherence (Elkilany et al., 2019). Anodizing improves the near-surface crystal structure and surface microscopic texture for superior mechanical performance as well as resistance to corrosion. Despite the fact that anodized aluminium has a stronger surface than bare aluminium, it has a low to moderate wear resistance. The wear resistance of treated aluminium material is improved by increasing the thickness of the anodic coatings, which is generally significantly stronger than metal plating and most paint types but also more brittle. As a result, anodic films are less prone to cracking and peeling caused by ageing and wear but have a high tendency to crack due to thermal stress (Abdel-Gawad et al., 2019). Humans' ability to construct materials such as alloys has progressed from simple to complex compositions as technology improves, resulting in increased functionalities and performances, as well as the advancement of human civilization.

To improve the thermal, mechanical, and electrochemical properties of engineering materials, a variety of alloying elements have been utilized to reinforce aluminium alloy (matrix). Recent developments in stir casting, on the other hand, have reinforced the inherent microstructural properties as well as many other desirable properties, such as outstanding corrosion resistance, high hardness and strength, fatigue, fracture, thermal stability, and irradiation resistance, in such a way that they outperform traditional alloys with a variety of potential applications (Zhang et al., 2018). A recent advancement in the stir casting method has resulted in a two-step mixing procedure known as double stir casting. The technique involves heating the matrix material over its liquidus temperature, then allowing the liquid metal to cool to a semi-solid condition between the liquidus and solidus temperatures. When the semi-solid condition is reached, the preheated reinforcement particles are inserted into the melted matrix material. The matrix and reinforcement combination is once again heated to a fully liquid condition and properly mixed as needed. When opposed to ordinary stirring, the two-step mixing procedure in double stir casting has the advantage of breaking the gas layer surrounding the particle surface, which would otherwise obstruct good fluidity between the particles and the molten metal. As a result of the abrasive action in the semi-solid state caused by the high melt viscosity, the gas layer is broken during particle mixing. Su et al. (2012) developed a three-step stir casting method for nanoparticle-reinforced composite production. To disrupt the early clumping of microparticles, the reinforcement and Al particles are first combined in ball mills.

After that, mechanical stirring is used to mix the composite powder into the melt. To increase the dispersion of reinforced particles, the composite slurry is sonicated with an ultrasonic probe or transducer after proper stirring. Kumar et al. (2021) used a three-phase induction motor to electromagnetically stir the aluminium melt and found that a small grain size structure improved particle–matrix interface bonding. Particulate reinforcement is uniformly distributed into the aluminium melt, which is mixed using a mechanical device such as a gearmotor, depending on the requirements. Stir casting presents a problem in terms of reinforcing particle segregation due to particle settling during solidification. The particle distribution in the ultimate solid state, the wetting condition of the particles with the melt, the pace of solidification, and relative density are all determined by the mixing strength. How well dispersed the reinforcement particles are in the molten mix is determined by the mechanical stirrer design, stirrer position in the melt, melt temperature, and the characteristics of the reinforcement particles.

7.2 ADDRESSING PACKAGING INDUSTRIAL MATERIAL CHALLENGE

Most research arose from a desire to address production/materials issues encountered, especially in the packaging industry as a result of the varying mechanical capabilities, thermal instability, and unequal microstructure of commonly used aluminium alloys. The mechanical performance of the alloyed material is improved by adding appropriate alloying components, organic or inorganic composite using the stir casting method. When compared to alternative techniques of reinforcing alloying materials, the stir casting method has the advantage of mass production. When it comes to mass production of MMCs, this approach is chosen because of its economic advantages. The capacity to adjust physical parameters such as thermal expansion, thermal diffusivity, density, and mechanical properties such as tensile and compressive strength, tribological behaviour, creep, and others by varying the filler phase has recently piqued the interest of researchers. In addition, due to the increased demand for aluminium matrix composites (AMCs) in industrial applications due to their unique mechanical, material, and tribological properties, considerable advancement in this area has occurred.

7.2.1 Alloys and Series Designation

Because aluminium in its pure form lacks useful properties, it is usual practice to combine it with other elements to create an alloy (Varshney & Kumar, 2021). This alloying element's incorporation prior to casting enhances desirable features or characteristics in both primary and secondary aluminium to achieve a desired customer alloy specification. In recent years, a lot of effort has gone into designing and testing new alloy constituents in order to attain desirable qualities including corrosion resistance, machinability, and ductility. Individual waste material's constituent elements are determined by the features and properties originally required for the original product. Thus, recycling establishments' gathering and sorting of scraps into various alloys or series is critical to the success of secondary aluminium alloy production.

The complexity of the foundry and the quality of scrap infeed used are determining factors in selecting the melt procedures to be used; the process can either be to melt down and recast the aluminium of a specific composition, or it can be to melt multiple alloy types in one batch to achieve a new specification with the inclusion of other elements, which are sometimes recycled. Anderson et al. (2018) present a detailed guide on identifying aluminium alloys based on their composition, heat treatment, and temper.

7.2.2 Cast Vs. Wrought Alloys

The first step in locating alloys is to categorize them as either wrought or cast alloys. The ductile features of wrought aluminium alloys are due to the reduced percentage content of elements in the alloy that improve cold working properties, such as when rolled, forged, pressed, or extruded. A wrought product could be a rolled aluminium sheet or an extruded window frame part. Cast alloys are alloys with a high degree of formability that can be cast into shape and finished using a variety of machining techniques. The main distinction between wrought and cast alloys is the percentage content of the alloying elements contained in each, yet they both contain the same alloying elements as required for the cast product's intended qualities. An aluminium engine block is a good example of a final cast product (Brough & Jouhara, 2020). The international alloy designation system is a generally acknowledged nomenclature system for wrought alloys, whereas the Aluminium Association provides chemical compositions and nominal densities for wrought alloys. To make identifying the composition and alloying constituents easier, wrought alloy is described using a four-digit method. The predominant alloying element is indicated by the first of the four numbers. Depending on the location, different designation schemes are used for cast aluminium alloys. The Aluminium Association has produced a simple chart for this purpose. The numbering system consists of four digits with a decimal point (0, 1, or 2), where xxx.0 refers to a casting and xxx.1/2 to an ingot. The first digit represents the principal constituent alloying elements when compared to the wrought alloys above. The second and third numbers specify the aluminium's predicted minimum composition. The alphabet is added to the numbering system to identify tempering techniques in the form xxxx-x, with the final x being: F (no treatment), H (strain hardened), O (completely annealed), T (tempered by heat treatment), W (solution heat treated) (Caetano, 2017).

7.2.3 Conceptualized Review on Composite Reinforcement Using Stir Casting

Researchers have previously constructed a few types of aluminium composites using organic and inorganic reinforcements to achieve a set of desired mechanical properties. Amit et al. (2020) investigated the impact of B4C and waste porcelain ceramic particle reinforcements on the mechanical and tribological properties of a high-strength AA7075-based hybrid composite. The authors used AA7075 alloy (99% pure) with a density of 2.8 g/cm^3 as the matrix, with B4C powder with a density of 2.52 g/cm^3 and an average particle size of 50 nm as the reinforcement particles, and porcelain powder containing 34% Kaolin, 14% ball clay, 31% feldspar, and 21% quartz as the reinforcement particles. A ball mill working at 200 rpm for 40 minutes ground the porcelain used in this study into powder. The powder had a particle size of 53 nm and a density of 2.4 g/cm^3. Furnace components and a graphite stirrer blade make up the experimental setup. The liquid state stir casting procedure was used to create the B4C porcelain-reinforced AMCs. To begin the fabrication process, both the AA7075 alloy rods and the reinforcing particles (B4C and porcelain powders) were heated individually in a muffle furnace for two hours at a temperature of 2000°C to drain moisture from the materials. In a graphite crucible at the furnace's base, the AA7075 alloy was totally melted at temperatures between 850°C and 200°C. The AA7075 was charged with a weighted amount of preheated reinforcing materials (B4C and porcelain). At 9300°C, a two-blade stirrer was used to swirl the molten mix for ten minutes at a constant speed of 350 rpm. To improve the wettability of the material, 1% magnesium was added to the mix.

Because the molten mixture was poured into a rectangular graphite mould before thorough stirring, the composite samples were sliced using a wire cut electro discharge machine and then made ready for the purpose of mechanical analysis. Various mechanical characterizations performed on the produced composite samples revealed that B4C particles were dispersed satisfactorily throughout the matrix. The density of the samples generated is said to decrease as the percentage weight of porcelain increases. The author also found a drop in porosity percentage from AMC-1 to AMC-4 hybrid composite, from 0.833% to 0.64%, however uniform distribution of reinforcement particles in the matrix was thought to be the reason for this incremental figure. In addition, due to an increase in the weight percentage of porcelain, AMC-5 showed an increase in percentage porosity. As the percentage weight of porcelain, B4C, increases, the hybrid composite's ultimate tensile strength grows linearly. The interfacial bonding between the aluminium matrix and the reinforcement (porcelain, B4C) particles, as well as load transfer between the two, was found to be responsible for the sample composites' optimal tensile strength. The compressive strength of an unreinforced AA7075 was found to be much lower than that of reinforced samples with different weight percentages (4, 8, and 12 wt%), with a severe decline in compressive strength at percentage weight 16 wt% porcelain. When comparing the unreinforced matrix to samples, the sliding wear study revealed that the unreinforced matrix had greater wear loss (AMC-2 to AMC-5). However, adding more porcelain to the composite sample than 16 wt% had a negative impact on its wear resistance. The author additionally subjected samples to loads of 10, 20, 30, and 40 N in order to determine the friction coefficient.

The addition of porcelain improves the anti-friction properties of the base matrix; however, excessive wear loss and friction coefficient were detected at the start of the 10 N load application, but these values were reduced to a stable value as the pin surface was meshed with the counterface. Furthermore, when the load was increased to 40 N, the coefficient of friction (COF) was reported to decrease. The sliding wear of samples was tested, and it was discovered that for all sliding ranges, porcelain-free samples have larger wear loss than porcelain-reinforced samples. The AMC-4 with 12 wt% porcelain and 4 wt% B4C composite, on the other hand, had the best wear resistance, which could be attributable to the wettability of the reinforcement material and matrix alloy. The addition of porcelain to a concentration of 16 wt%, as in AMC-5, resulted in poor wear resistance. Huda et al. (2019) studied the mechanical and wear properties of Al_2O_3 nanoparticle-reinforced AA7075 AMCs. The stir casting method was used to create the manufactured composite, which used AA7075 as the matrix and Al_2O_3 as the reinforcing particles. Matrix material (AA7075) was first cut into 1–2 cm^3 cubes, and then washed many times with alcohol and distilled water. The easily cleaned AA7075 cubes were dried in an electric heater at a temperature of 2000°C after being exposed to hot air at a temperature of 1000°C. Before aluminium and argon gas were charged into the oven, accumulated air within the furnace was vacuumed out; nonetheless, the oven temperature was raised to 8500°C. As a result, the reinforcement particles were injected into the molten matrix alloy using a gas pump. For the produced composite, the author employed the following process parameters: 450 rpm, 8500°C stirring temperature, 2000°C preheat temperature of reinforcement, 10–12 minutes stirring duration, 300°C–400°C preheated temperature of the metal die-casting mould, and 1 g/kg min powder feed rate.

The molten matrix was stirred for four minutes at a gradual speed increase to 450 rpm. On AA7075 and composite samples with different percentage weights of Al_2O_3 (1 w%, 3 w%, and 5 w% Al_2O_3), microstructural and mechanical analyses were performed. The results showed that the sample composite has a higher yield and ultimate tensile strength, which increases as the percentage weight of reinforcing particles increases. It was discovered that a composite sample containing 5% Al_2O_3 had the best yield and tensile strength. The increased strength of the samples was also attributed to heterogeneous nucleation sites formed during solidification as a result of nanoparticle inclusion. The inclusion/increase in nanoparticles results in a considerable increase in composite hardness, according to a hardness test of composite samples. The homogeneous distribution of reinforcing particles in the composite samples was credited with the high hardness. Because the hardness properties of Al_2O_3 are easily transferred to the matrix, the sample containing 5 wt% Al_2O_3 displayed the maximum hardness. To investigate the pace at which material wear occurs, composite samples were submitted to a pin-on-disc tester machine. When subjected to stresses ranging from 10 to 50 N, the sample containing 1 wt% Al_2O_3 showed the maximum wear rate.

The author found that as the number of reinforcement particles grows, the wear rate of the AMC decreases; however, even with more reinforcement particles, the wear rate can increase for loads of 30 N and higher. The research on composite wear in relation to sliding distances revealed that as the sliding distance increases, the wear rate increases proportionally. This, according to the author, is due to the incorporation of Al_2O_3 particles in composite surfaces, which act as acute asperities. The surface area between the disk and the test specimen is reduced as Al_2O_3 particles protrude at tiny distances. Increased sliding distance, on the other hand, creates abrasion between the sliding surfaces, which may be the source of the observed wear rate rise. Alaneme and Bamike (2018) investigated the mechanical and wear properties of aluminium composites reinforced with quarry dust and silicon carbide. The authors used SiC and quarry dust as reinforcement particles in double stir casting to create an Al–Mg–Si alloy (AA6063) composite. At an 8 wt% reinforced Al–Mg–Si matrix composite, the ratio of SiC to quarry used for sample production was 1:0, 3:1, 1:1, and 0:1. The SiC particle size in the samples was 28 μm, and the quarry dust was sieved to a mesh size of less than 50 μm. In a laboratory oven, both reinforcement particles were warmed independently for an hour at 250°C. In a gas-fixed crucible, the matrix alloy was totally melted to a temperature of 750 ± 30°C above the liquidus temperature of Al–Mg–Si alloy. The first melting process was preceded by a heat loss of melted alloy to 600°C, primarily to preserve a semi-sold state. The semi-solid molten aluminium alloy was manually swirled for two minutes before adding preheated quarry dust and SiC particles to obtain a homogeneous dispersion composite mix. As a result, the generated semi-solid composite was extensively agitated at a speed of 400 rpm for ten minutes at an enhanced temperature of 800°C ± 30°C before being put into sand moulds.

In order to decrease incipient voids, porosity, and blow holes that may occur during casting, the produced as-cast composites were treated to cold rolling and heat treatment methods. The authors machined the composite samples to a diameter of 15 mm and a length of 20 cm, following which they were cold rolled with round grooves to a diameter of 13.5 mm. As a result, the rolled samples were heated to 350°C for one hour before being quenched in water to relieve the tensions induced inside during cold rolling. The composite samples were characterized mechanically for hardness, tensile strength, fracture toughness, and wear characteristics. The authors found that as the wt% of quarry dust in composites increases, the hardness attributes of the composites decrease marginally. Samples A2, A3, A4, and A5 with 1:3, 1:1, 3:1, and 1:0 wt ratios of quarry dust to SiC, respectively, showed a reduction of 0.5%, 2%, 2.3%, and 5% when compared to composite A1 with 100% SiC. Quarry dust was shown to have no negative effects on hardness characteristics and can be used as a substitute for SiC. Based on the composite samples' responses to tensile strength, specific strength, and strain to fracture tests, the authors concluded that there was no consistent variation in ultimate tensile strength, with all recorded values falling within a 10% range; however, sample A3 with a reinforcement ratio of 1:1 had the highest tensile strength. Due to the higher modulus of elasticity of SiC compared to the principal constituent of quarry dust, sample 1 with a single SiC-reinforced composite was expected to have the maximum tensile strength. According to the authors, quarry dust can be a good substitute for SiC as reinforcement particles in applications where tensile strength is critical. Furthermore, the specific strength of the composite samples was found to follow the same pattern as the tensile strength. However, the same strain-to-fracture results show that adding quarry dust to composite samples significantly increases strain to fracture.

As a result, an increase in quarry dust has been shown to improve ductility and plastic strain without fracture. The fracture toughness of samples improves with a weight per cent increase of quarry dust. The presence of brittle and relatively tougher SiC particles in composites was reduced, resulting in increased fracture toughness. The authors go on to say that when quarry dust was used as a reinforcement particle in aluminium, it increased the material's fracture resistance more than SiC. The wear index of composite samples improved marginally with an increase in quarry dust weight per cent less than a 7% drop in wear resistance, indicating that adding quarry dust as reinforcement to Al–Mg–Si alloy composites will not significantly diminish wear resistance. The microstructure of composite samples was examined using a

Zeiss optical microscope with the ability to analyze microstructural pictures. Samples were sufficiently polished and then etched using a solution of 25 mL methanol, 25 mL hydrochloric acid, 25 nitric acids, and a drop of hydrofluoric acid prior to microstructural investigation. Amouri et al. (2016) examined the microstructure and mechanical properties of Al-nano/micro-SiC composites made by stir casting. The authors employed A356 aluminium ingots as the matrix, with SiC powder in nano and micro sizes as reinforcing particles. The fabrication procedure in this study begins with the melting of A356 aluminium ingots in a graphite crucible at 800°C using a resistive furnace. 1 wt% Mg was added to the molten mix to control gas entrapment and wettability issues and then covered with a coverall. The SiC powders were warmed at 800°C for an hour before being added to the molten mix to extract moisture and increase the wettability of reinforcement particles with the molten matrix. Mechanically stirring the warmed reinforcement particles with molten matrix at 600 rpm for seven minutes. To lessen oxidation during stirring, nitrogen was dropped into the molten mix surface. After that, the composite is poured into a preheated cylindrical mould.

The same method was utilized to create NMMCs with micro SiC concentrations of 0.5 and 1.5 wt%. The authors used T6 heat treatment on the composite samples to see how it affected microstructural and mechanical qualities. After heat treating the samples to 540°C and keeping them at that temperature for six hours, they were water quenched to room temperature. Following that, samples were artificially aged at 190°C for four hours before being air-cooled. Both cast and heat-treated samples were made using metallographic procedures and etched Keller's reagent before microstructural and mechanical tests (3 mL HCl, 2 mL HF, 5 mL HNO_3, 190 mL H_2O). The researchers observed that SiC particles impact dendritic formation decrease utilizing a sample's microstructural image analyzer. The length of dendrites formed decreases as the weight per cent of nano SiC or micro SiC particles increases. Micro-SiC particles refine dendrites better than nano-sized SiC particles, according to the authors. The rise in the number of growing dendrites was linked to the reduction in the timeframe within which solute atoms were discharged into the surroundings. Pure A356 and micro/nano MMCs were put through tensile and compression tests.

With the addition of SiC particles, the strength of MMCs is greatly boosted at the expense of ductility, according to the authors. The composite containing 1.5 wt% nano SiC exhibited the best tensile and compression strength. The interaction of well-dispersed micro/nano SiC particles with matrix dislocations was found to be the main source of strength. The nano-sized SiC particles, on the other hand, were more evenly distributed in the molten mix than the micro-sized SiC particles. The poor ductility of pure A356 was linked to particle size, weight fraction, and spacing; as a result, an increase in particle average size reduces the driving stress of a given volume fraction. It was also discovered that numerous Mg_2Si particles settled during the T6 treatment's ageing stage, forming the foundation for MMC's greatest strength of 1.5 wt% nano-SiC. Due to the combined effect of precipitation hardening of Mg_2Si and nano-sized SiC particles on ductility, the lowest fraction strain (ductility) was recorded in the same MMC containing 1.5 wt% nano SiC. SEM images of cracked surfaces of samples subjected to tensile testing reveal the existence of dendritic structure, supporting the theory that fractures start in the eutectic zone and plate cracks such as Si. T6 thermal treatment of samples is said to result in a less than 40% reduction in dendritic structure in the fracture site. It has been found that increasing the concentration of nano SiC particles has no effect on the structural alteration of the A356 matrix, as the same fracture surface was observed in matrix compositions including 0.5 wt% nano SiC. The mechanical evaluation further demonstrated that A356 with 0.5 wt% nano SiC and cast-T6 treated A356 have nearly identical stress–strain capabilities. However, in MMCs containing 1.5 wt% nano SiC and 5 wt% micro SiC, dispersion of SiC particles favourably promotes a totally cleavage fracture with a granular flat surface.

Adat et al. (2015) used a stir casting route in reinforcing Al356 alloy with Al_2O_3 and fly ash particles at different percentages (2%, 4%, and 6%). At 400°C in the furnace, the required percentage weight of A356 alloy ingots was charged into a cast iron crucible melt. 1% pure Mg was added to the charge at 600°C in a semisolid state to improve wettability. The substance was then allowed to cool for an hour, essentially to achieve a temperature of 800°C. Scum powder (0.05% Al356) was charged into the melt at a percentage weight that promoted the presence of impurities on the liquid melt's surface. Hexachloroethane was used to eliminate the presence of gas in the molten alloy (0.05% of Al356). Before charging into the Al356 melt, reinforcement particles were warmed to a temperature of 300°C–400°C for an hour. At a rate of 0.5 g/s, the Al_2O_3 alloy was poured into the vortex from the centre. The stirrer mechanism swirled the melt at 300 rpm for 7–10 minutes while moving it down slowly from top to bottom with enough clearance from the bottom. The liquid was poured at a temperature of roughly 800°C. During the pouring of liquid composite into an MS permanent mould for melt fluidity, a constant pouring rate and distance were maintained. Before pouring began, a rectangular metal mould with dimensions of 220 × 220 × 20 $(mm)^3$ was preheated for an hour. The stir casting procedure revealed that the generated AMMCs have a homogeneous distribution of Al_2O_3 and fly ash particles. The material soundness of the cast was increased as a result of preheating the mould prior to pouring. The addition of magnesium to the melt improved the wettability of the matrix and reinforcement. Except for 8% fly ash reinforcement, the percentage porosity in the cast composite increased linearly with particle weight fraction. The authors also discovered that composite materials reinforced with fly ash are less dense than those reinforced with alumina. The effect of fly ash reinforcement particles on composites, on the other hand, is visible in the decrease in composite density as reinforcement increases, whereas the density of composites reinforced with alumina grows linearly. The compression strength of all reinforcements is greatly increased; however, composites reinforced with fly ash have a lower compression strength than those reinforced with alumina.

Prakash and Jaswin (2015) used a stir casting procedure to create a composite with SiC and B4C particles as reinforcements and Al6061 as the matrix. The weight fractions of Al (90%), SiC (5%, 6.5%, and 8%), and B4C (5%, 3.5%, and 2%) were changed for particle addition. Heat treatment of aluminium rods was carried out at temperatures ranging from 850°C to 900°C for three to four hours, while SiC and B4C particles were melted at temperatures ranging from 840°C to 900°C for one to three hours, and then the surface of the reinforcing particles was oxidized. The ambient temperature of the furnace was originally elevated above liquidus temperature and then allowed to descend below liquidus temperature in order to keep the slurry in a semi-solid state in order to achieve a totally melted aluminium alloy. They advocated for manual mechanical mixing to overcome the homogeneity issue that arises when semi-solid composite aluminium alloys are mixed automatically. After the composite slurry had been reheated to a liquid state, an automatic mechanical mixing system was utilized to agitate it for ten minutes at a speed of 600 rpm. The results showed that the SiC and B4C particles were evenly scattered over all composites, as seen in the micrograph image. It was discovered that the temperature of the procedure has a direct impact on particle presence. AMMCs were created utilizing Al6063 as the matrix material and SiC particles with varying RVR (between 10% and 50% at 10% intervals) as reinforcement. Using the stir casting technique, a homogeneous mixture of matrix and reinforcement was created to produce cylindrical blank composites. The composites created have improved metallurgical characteristics, according to the findings. It was also discovered that an increase in the quantity of SiC reinforcement has a good effect on metallurgical properties only up to 40%wt SiC, after which there is greater clustering of the reinforcements.

The effect of fly ash hybrid reinforcement on the mechanical properties and density of aluminium 356 alloy was investigated by Kulkarni et al. (2016). As matrix and reinforcing materials, the author employed a blend of Al356 with fly ash, alumina, and a hybrid combination of both. The matrix composite was created using the stir casting method, with reinforcement particles made up of 4%, 8%, and 12% fly ash, 4%, 8%, and 12% alumina, and the equal amounts of fly ash and alumina in 4%, 8%, and 12%. With the use of a size shaker, a particle size of less than 100 m was attained. The reinforcement particles were injected into the molten matrix, which reached a liquidus condition at 800°C. In a crucible with a capacity of 4 kg, 3000 g of Al356 alloy and the equivalent per cent weight of reinforcement were melted. After reaching a liquidus temperature, the melt was held at that temperature for eight minutes while stirring at 500 rpm with a three-step nine-blade stirrer driven by an electromechanical drive. Before the reinforcement particles were charged, 1 wt% magnesium was added to the molten matrix, primarily to improve the matrix's wettability with the reinforcement particles. In addition, the author utilized 0.05% hexachloroethane to separate layers of scum from the melt's surface. After sufficient homogeneity was established, a molten composite was placed into a preheated round plate mould and allowed to cool at room temperature. With the exception of 8% reinforcement of fly ash, the generated composite showed an increase in porosity as the percentage weight fraction of reinforcement increased. When comparing density, it was discovered that composite materials reinforced with fly ash have a much lower density than those reinforced with alumina. When the percentage weight of fly ash is increased, the composite density decreases, and when the percentage weight of alumina is increased, the composite density increases; nevertheless, using equal amounts of fly ash and alumina keeps the composite density near to that of Al356 aluminium alloy. The results of the author's compression strength experiments revealed a linear relationship between compression strength and the percentage increase in reinforcing particles. When compared to composites reinforced with alumina, the compression strength of composites reinforced with fly ash is lower. As a result, according to the author, hybrid reinforcement keeps composite density near to that of an unreinforced matrix while encouraging better material qualities than fly ash-reinforced composite.

The authors further highlighted that the strength of composites is at the detriment of increased porosity. It was also reported that a reduction in strength of A356-8% alumina composites up to 156.86 MPa for 7% porosity; however, A356-8% fly ash composite exhibited better properties as compared to A356-8% alumina as a result of presence of porosity. Using the stir casting and compo casting methods, David et al. (2013) created an Al6061–fly ash–SiC composite. A mix of reinforcing particles with varying percentage weights of 7.5%wt, 10%wt for SIC, and 7.5%wt for fly ash was used. 1500 g of Al6061 rod was put into a coated graphite crucible and cooked in an electric furnace to 920°C. A mechanical stirrer was used to provide adequate churning of the molten matrix in order to promote the production of fine vortices. The molten aluminium matrix was charged with a continuous feed rate of warmed SiC and fly ash in the desired proportions as per the sample. Preheating of the reinforcements was done at 920°C for 90 minutes. To improve wettability between matrix and reinforcement particles, 1% magnesium particles were utilized. By adding argon gas to the mix, an inert atmosphere was created. To increase the dispersion of silicon carbide and fly ash particles in the molten matrix alloy, the author used two stages of stirring. However, the first step of stirring was carried out while the slurry was still semi-solid, and the second stage of stirring was carried out once the matrix melt had reached a liquidus temperature.

The author observed that the reinforced AA6061 alloy matrix demonstrated the existence of an aluminium dendritic network structure, which he attributed to the super cooling of the composite during modification. In the microstructure, Mg_2Si precipitate was also found. The reinforced composite with varying weight percentages of fly ash and SiC, on the other hand, had a homogenous distribution with no porosity or cracks in the samples. During the solidification of the AA6061–fly ash–SiC composite, fly ash and SiC were seen to separate from the mix in the direction of the refined aluminium grains. The presence of FA in the SiC performs also prevents the development of aluminium carbide (Al_4C_3), according to the author. The homogeneous distribution of FA and SiC particles observed can be attributed to a good stirring method and the application of the right process parameters. The micro and macro hardness of aluminium composite samples increased linearly with the increase in reinforcing weight per cent, according to the results of the hardness test. When reinforcement particles are added to a molten matrix, their surface area increases, causing the matrix grain sizes to shrink. The hard surface particles were shown to be responsible for the composite's resistance to plastic deformation. The combination of SiC and fly ash particles in the matrix improves composite strength and consequently increases composite resistance to tensile stresses, according to the author. Because the research only looked at a few combinations of different weight percentages, only a few composite samples were created.

Nair and Joshi (2015) produced an AMC utilizing the stir casting method. The base metal was Al6061, and the reinforcement particles were powdered silicon carbide (SiC) with a 320 mesh size. In a graphite crucible, an Al6061 sample was melted at 650°C. A 10% weight fraction of SiC powder with a 100 mesh size was charged into the molten matrix, along with a 1% weight fraction of magnesium powder to improve wettability as porosity decreased. The material used to stir the mixture was a 100 mm × 25 mm diameter graphite rod. The shaft's butt was externally threaded, and it was attached to a three-foot stainless-steel rod that was internally threaded in the middle. An electric gearmotor at 250 rpm was used to mechanically agitate the liquid for ten minutes before transferring it into the 110 × 110 × 15 mm mould. The aluminium matrix composite mixture in the mould was given enough time to solidify. The result reveals that silicon carbide particles are equally distributed throughout Al6061 and that reinforcement particles and matrix have a strong connection. The use of magnesium powder reduced porosity and enhanced wettability. This study simply looked at the development of a sample Al6061–SiC without changing any process factors like stirring speed, stirring temperature, reinforcement preheat temperature, stirring time, preheated mould temperature, powder feed rate, or stirrer blade design.

7.3 PROGRESSION ON STIR CASTING PARAMETERS

Many stir casting parameters, such as stirring time, stirring speed, stirring temperature, stirrer design, stirrer position, melt temperature, mould temperature, reinforcement preheat temperature, blade angle, particle sizes, and weight fraction, all contribute to a satisfactory fabrication of stir cast material that will deliver for the intended use. Rangrej et al. (2021) investigated the effect of stir casting parameters on reinforcement particle dispersion during the manufacture of MMCs. Al-alloy LM25 and SiC were employed as matrix and reinforcement particles in this study. The initial melting of the LM ingot in a graphite crucible at 750°C in a muffle furnace is the first step in the casting process.

In a muffle furnace, the reinforcement particles were preheated to a temperature range of 800°C for three hours, mostly to extract moisture. The preheated silicon carbide particles were added after the molten matrix had reached its liquid state, and the molten mix was thoroughly agitated using a stainless steel stirrer operated by a DC motor at speeds of 500, 600, and 700 rpm. The molten mix was agitated for ten minutes to ensure uniform reinforcement particle dispersion, and the process was completed by pouring the molten composite into a prepared mould. Both porosity and microstructural analyses were performed on the Al cast samples. The density and porosity of samples were tested at various speeds and blade angles. The presence of low-density material was shown to cause the theoretical density of composites to be higher than the measured density. The porosity of samples produced at 600 rpm and a 60° blade angle was reported to be low. On readily grinded and etched samples, a microstructural investigation was performed. Both the as-cast Al alloy (LM25) and the Al–SiC MMC micrographs demonstrate that the composite samples have homogeneous dispersion of SiC particles with traces of porosities. In comparison to Al-alloy, the grain size of Al–SiC composite was reduced by 10.5% using BIOVIS software. Micrographs revealed the region where porosity and particle clustering were concentrated. The authors concluded that reducing the size of the reinforcing grain improves the mechanical characteristics of composites.

Tamilanban and Ravikumar (2021) investigated the effect of steering speed on the mechanical characteristics of an Al–Mg–Cu composite reinforced with SiC that was cast using the stir casting method. To make an Al MMC with 12% SiC particles, 4% Mg, and 2% Cu content, the authors used a stir casting approach combined with a mechanical stirring arrangement. 1 kg of 6061 aluminium alloy was put into a graphite crucible as a matrix and melted at temperatures between 750°C and 800°C. The SiC particles were warmed to a temperature of 350°C, primarily to remove moisture and increase wettability. The hot SiC was carefully mixed with copper and magnesium metal powders in the exact amounts necessary. Thus, the preheated SiC particles were charged into the molten matrix while an impeller was used to mechanically swirl the molten mix at varied speeds of 500, 700, and 900 rpm for 12 minutes.

The properly mixed molten mixture was placed into a cast iron mould and allowed to cool until it solidified into a solid MMC casting. The authors also poured the molten mix into a steel mould of 120 mm length, 120 mm width, and 25 mm thick, preheated at 350°C. Cleaning and machining of samples were completed prior to microstructural testing and mechanical characterization. Tensile, hardness, and wear testing were among the mechanical tests performed. The authors used Wilson Wolpert Hardness Equipment to measure samples' hardness at three separate places, subjecting test samples to a 0.5 kg force for a 15 s dwell duration. The samples were tensile tested using a universal testing equipment according to ASTM E8. Samples were polished with a fine emery cloth before being immersed in paraffin (kerosene) oil for the wear test. The pin-on-disk wear test equipment was employed in a dry sliding condition at room temperature, and the COF of the samples was determined using a DUCOM model wear test machine. During the wear test, a test load of 40 N, a sliding speed of 0.393 m/s, and a sliding distance of 1000 m were maintained. Optical microscopy (OM) and SEM were used to examine the microstructure of the readily grinded, polished, and chemically etched samples using Keller's reagent solution. The homogenous distribution of SiC reinforcement particles has a direct link with stirring speed, according to the authors. The SiC reinforcement particles, as well as Cu and Mg powder, are equally dispersed in the Al MMC produced at a stirring speed of 700 rpm. With the addition of Cu and Mg powder, however, dendritic structures were detected in samples swirled at 900 rpm; consequently, SiC particles agglomerated in the interior portion while few particles were visible.

Reinforcement particles and alloying components were also found to cluster in a specific direction. As the Mg content rises, so does the homogeneity of reinforcement particles with the matrix, which improves wettability. The authors tested yield strength and ultimate tensile strength on test pieces manufactured at three different speeds. When compared to samples produced at different speeds, it was shown that Al MMC fabricated at 700 rpm stir speed had better mechanical properties. The superior mechanical properties of the sample manufactured at 700 rpm stirring speed were attributable to the distribution of load SiC particles from the Al matrix and dislocation of induced density strengthening of the MMC from the reinforcement particles, according to the authors. Better SiC particle alignment was also related to the increased mechanical characteristics of samples manufactured at 700 rpm stirring speed. However, particle resistance on dislocation slip resulted in a considerable drop in material elongation. Mg and Cu inclusion increased the quality of the samples by directly improving solid solution strengthening and wettability of SiC reinforcement particles with the Al matrix. However, caution was exercised in guaranteeing a 4%wt Mg inclusion, while precipitation hardening was blamed for the increase in strength and hardness properties caused by Cu inclusion. The hardness qualities of samples manufactured at three different rates are also compared by the authors. The addition of Mg and Cu to Al matrix composites has been shown to increase material hardness.

Furthermore, as the stirring speed is reduced, the hardness of composite samples increases. This finding was thought to be the result of a high-speed repelling force (push) on reinforcement particles towards the interior region. As a result of the presence of SiC particles at the exterior region, samples manufactured at lower stirring speeds had better toughness. There was also a difference in hardness qualities at different places of the sample manufactured with 900 rpm stirring speed compared to samples fabricated with lower stirring speed. The wear rate/COF of composite samples was also investigated. It has been found that adding SiC reinforcement particles to MMC reduces the wear rate. When compared to other samples, samples prepared at 700 rpm stirring speed had a high COF value. In composite samples manufactured at 700 rpm stirring speed, a homogeneous SiC distribution was obtained, resulting in a composite surface with a higher COF. The dispersion of Cu, Mg, and SiC reinforcement particles along the surface of MMC samples was visible using SEM microscopy. The influence of various process parameters on microstructural behaviour and mechanical qualities of aluminium (Al + 5% Zn + 0.1% Sn) using commercially available manganese oxide (concentration powder) as reinforcement particles was explored by Abdelrahman et al. (2020). In his investigation, the authors apply the various process parameters provided in Table 2.1. It was discovered that the best material integration and dispersion occurred at a stirring speed of 700 rpm, as opposed to 300 and 1100 rpm. The vortex created at 300 rpm was not powerful enough to trap the reinforcement into the molten matrix for a more uniform matrix-reinforcement mixture, resulting in reinforcement particles accumulating on the melt surface. However, stirring at 1100 rpm causes a forceful development of vortex that extends to the lowest section of the molten matrix, resulting in severe air trapping and oxide formation.

Kumar et al. (2019) used the stir casting approach to examine the effect of casting parameters on particle distribution in MMCs. The authors employed an aluminium alloy (Al8011) as the matrix and silicon carbide (SiC) as the reinforcing material. Taguchi design expert was used to create an MMC with eight samples and two levels in this study. Volume concentrations of 5%, 10%, and 10% SiC particles, 1.24 MPa-s and 1.04 MPa-s at 700°C and 800°C working temperatures, 45° and 90° blade angles, 40% and 20% impeller placement from crucible base, and 10 and 15 minutes of holding time, respectively, were used in this investigation.

Prior to the casting of samples, an electric melting furnace was prepared to a temperature of 800°C, and then Al8011 aluminium alloy billets were charged into the furnace using a crucible while the heating continued. The reinforcement particles were preheated in an electric oven set to 350°C. Prior to melting aluminium billets, quickly warmed reinforcing particles are charged into the molten matrix. As per the samples, the addition of reinforcement particles into the molten matrix was done simultaneously with continuous stirring for 10–15 minutes. As a result, the molten composite matrix was poured into a die to cast a sample. The structural investigation of the developed samples was performed using OM and SEM. With the help of a wire-cut electric discharge machine, samples for examination were chopped into 10 × 10 × 10 mm^3 sizes. Micrograph images of samples 1 and 3 reveal a moderate dispersion of reinforcement particles in the mix; for sample 2, SiC distribution was uniform across the sample, with fine and poor distribution in some areas.

For samples 5 and 4, a moderate and fine distribution of SiC was observed at nearly all sites. The microscopic images of samples 6 and 7 exhibit a moderate and poor distribution of SiC, respectively, but the existence of grain formation was observed with a well-distributed particle distribution. The lack of large pore and particle clustering in micrographs of samples 2, 4, and 8 were attributed to the holding period, which favours uniform dispersion in the molten mix as well as efficient interface bonding between reinforcement particles and matrix. The authors also report that uniform particle dispersion was achieved by placing the impeller 40% closer to the base of the crucible, which favours fine vortices and prevents air bubbles from becoming trapped in the Al matrix. The 45° blade-angled stirrer was credited with the best shearing rate and vortex depth. However, due to turbulent vortex generation, a higher blade angle, such as 90°, promotes air bubble entrapment in the molten matrix, resulting in particle clustering in the matrix. Pores and particle clustering are aided by a greater processing temperature combined with a higher blade angle, according to the authors. However, higher processing temperature and stronger blade angle resulted in higher particle clustering despite longer holding times and lower impeller positions. Mechanical characterization of samples 2, 4, and 8 included tensile strength, hardness, and a tribological (wear) test. The ultimate tensile strength, yield strength, and per cent elongation of the samples were examined using a universal testing machine, whereas the ultimate tensile strength, yield strength, and percentage elongation of the samples were checked using an ASTME-8 standard (H50KL, Tinios Olsen Computerized model). For tensile strength, the average result of the mechanical characteristics for separate samples was computed, and the maximum average tensile strength for sample 8, sample 4, and sample 2 was 69.45, 67.3, and 61.55 MPa, respectively. In comparison to samples 4 and 2, sample 8 with 10% SiC volume concentration, 4° blade angle, 800°C melting temperature of Al8011, stirrer position of 40% from crucible base, and holding period of ten minutes has remarkable tensile strength.

The agglomeration of SiC particles in a composite increases the material's tensile stress resistance. As a result, a sample with a 10% volume percentage of SiC particles has a high tensile strength. A Brinell macro-hardness tester with a 15 kg load was used to test the samples' hardness for a duration of ten seconds. The average of three readings taken at various points was determined. With the identical settings as before, the hardness test for sample 8 yields an increased hardness result of 83BHN. As a result, the authors found that the hardness of composites increases as the volume proportion of SiC increases. It was also stated that areas with accumulated particles have a higher hardness, whilst areas with less particles have a lower hardness.

Pin-on-disc wear testing equipment was used to conduct friction and wear tests on the samples (TR-20LE, DUCOM). 6 mm diameter by 25 mm long test samples were machined and tested in air according to the ASTM G-99 standard. For this test, a hardened steel with a hardness of RC60 was machined to a diameter of 55 mm, a thickness of 5.5 mm, and a thread size of M6. With the use of an electronic weighing equipment with a precision of 0.001 g, the before and after weight of samples were used to quantify material loss. Due to the presence of a high volume proportion of SiC particles, the authors stated that sample 8 with the identical process conditions as before has the lowest wear rate and weight loss.

The effect of process factors on the wear behaviour of copper-coated short steel fibres reinforced in LM13 aluminium alloy composites was investigated by (Chelladurai et al., 2018). For the development, the authors used a three-factor, five-level central composite experimental design with a variety of process parameters such as weight per cent of reinforcement, stirrer speed, and pouring temperature. The matrix was made of LM13 aluminium alloy, while the reinforcement components were commercially available steel fibres with a diameter of 133 µm. The authors used a stir casting set-up to melt 1 kg of LM13 aluminium alloy at a temperature above its liquidus state and then used hexachloroethane tablets to degasify the material. A fixed amount of copper-coated steel fibres was preheated to 200°C and gradually dosed into the matrix melt while stirring continued, according to the experimental design. The set-up consisted of a stainless-steel stirrer that rotated at a steady pace. An H11 steel die mould was prepared to a constant temperature of 225°C using a ceramic electric heater prior to pouring the molten composite; following that, the progressive bottom pouring method began. As a result, a cylindrical composite casting with a diameter of 50 mm and a height of 130 mm was created.

The microstructural study of samples manufactured with 5.5 wt% copper-coated steel fibres at 750°C pouring temperature at varying speeds was explored. Micrograph analysis revealed that samples manufactured at a 350 rpm stirring speed had agglomerated steel fibres and a non-uniform dispersion of LM13 aluminium alloy. Samples manufactured at a stirrer speed of 800 rpm showed partial dispersion of copper-coated steel in LM13 aluminium alloy, whereas samples produced at a stirrer speed of 575 rpm showed uniform distribution of copper-coated steel fibres. As a result, it is clear that a high stirring speed favors enough vortex formation, and the dispersion of reinforcing particles, resulting in improved mechanical properties. Dry sliding wear behaviour with a velocity of 1.7 m/s was one of the mechanical characterizations performed on samples, and a load of 10 N was applied to the sample at a sliding distance of 2000 m. As per design expert version 10 with a regression equation constructed in forecasting the weight loss, four samples were analyzed for average loss with respect to pouring temperature, stirrer speed, and weight per cent.

At all pouring temperatures and stirrer speeds, the authors found that increasing the weight per cent of reinforcement reduces composite weight loss. It was also discovered that increasing the stirrer speed from 441 to 708 rpm resulted in a significant reduction in composite weight loss at all levels of reinforcing particle weight per cent and pouring temperature. The weight loss reduction was ascribed to the consistent dispersion of copper-coated steel fibres in the LM13 aluminium matrix, according to the authors. Coagulation of steel fibres in LM13 aluminium alloy was also attributed to low stirrer speed; however, stirrer speeds greater than 600 rpm favourably enhance steel fibre dispersion in molten matrix. It was also discovered that pouring at a higher temperature while stirring at a slower pace causes composite weight reduction. However, a higher pouring temperature combined with a faster stirring speed accelerates composite weight decrease. Worn surface analysis of the composite was performed, and it was discovered that the sample with 2.8 wt% reinforcement had deep continuous grooves on the surface parallel to the sliding direction, as well as delamination in nearly all places. Fine grooves and minute shallow grooves were visible on the surface and in some parts of the composite sample with 5.5 wt% reinforcing particles. Furthermore, fine grooves were observed on the surface of the composite with 8.2 wt% parallel to the sliding direction, despite the absence of delamination; this is due to the greater weight per cent of reinforcing particles (hard-coated steel fibres). Mondiu et al. (2017) used stir casting to investigate the microstructural behaviour and mechanical properties of aluminium alloy reinforced with Rice Husk Ash (RHA) composite. The hardness of Al alloy composites increases with increasing percentage weight fraction and stirring duration, but decreases with increasing reinforcing particle size, according to the author. As the weight fraction increases, composite hardness increases with particle size 150 μm; hence, the greatest composite hardness is seen at particle size 150 μm and 15% weight fraction. According to the author, the hardness of a reinforced composite sample was influenced, and at a high value with a stirring time of 30 minutes, particle size of 300 μm, and a weight fraction of 15%, hardness decreased for both high and low speed with an increase in particle size, while at a weight fraction of 10%, hardness decreased for both high and low speed.

Arulraj et al. (2017) used the Taguchi approach to create and analyze the influence of improving process parameters in stir casting of a hybrid metal matrix (LM25/SiC/B4C) composite. The author made a hollow cylindrical sample with a diameter of 50 mm and a height of 200 mm. In a crucible furnace, the matrix LM25 ingots were charged and melted at the appropriate temperature for material fluidity. Following the molten melt was gasified, it was continuously swirled at 500 rpm for ten minutes with an electric motor, after which reinforcement particles (SiC and B4C) were gradually injected into the melt vortex for good dispersion. A stir speed of 800 rpm, a stirring period of five minutes, and a preheated temperature of reinforcement particles at 400°C are the best casting parameters for stir casting of LM 25/SiC/B4C hybrid composites, according to the author.

Adebayo et al. (2017) focused on the impact of stirring speed on the microstructural behaviour and mechanical properties of an aluminium alloy (Al6061) reinforced with 15% silicon carbide particles (SiCp). According to the author, SiC clusters can be seen in some areas, and porosity can be detected at 300 rpm. Furthermore, a uniform distribution of SiCp in the aluminium was not accomplished; thus, 300 rpm is insufficient for 500 rpm, and reinforcement particle distribution is improved. However, at this speed, the vortex increases air trapping in the molten mix, resulting in casting flaws. They go on to say that a violent vortex created at a higher speed of 700 rpm induces the presence of oxide skins, gases, and impurities in the melt, which is detrimental to optimal particle dispersion. Using the stir casting technique, Ajagol et al. (2018) investigated the effect of SiC volume per cent and particle size on microstructure characteristics and mechanical properties of aluminium matrix. The authors made Al–SiC composites with SiC weight fractions of 5%, 10%, and 15%, respectively. The tensile strength and hardness of the generated samples were boosted by

SiC inclusion, according to Ajagol et al. (2018), whereas clustering of silicon carbide particles was seen in the aluminium matrix with increased porosity as SiC increased. According to the findings from the literature, improper or incorrect selection of aluminium matrix alloy and reinforcement particles, as well as incorrect varying of process parameters, will result in poor composite material with deficient chemical and mechanical properties, rendering the developed composite matrix unsuitable for its intended use. When constructing an AMC with substantial mechanical qualities, it is critical to keep this in mind. Stir casting technique is one of the regimes or processes used for creating aluminium matrix composites. The technology is unquestionably a cost-effective approach to mass-producing technical components for industries such as aluminium packaging, cookware, aeronautics, and automobiles. Through stir casting, many researchers have made significant contributions to the creation of AMCs. Stir casting parameters such as stirring time, stirring speed, stirring temperature, stirrer design, stirrer position, melt temperature, mould temperature, reinforcement preheat temperature, blade angle, particle sizes, and weight fraction have a significant impact on microstructural behaviour, chemical, and mechanical properties of aluminium composite, according to the researchers. They reported that reinforced aluminium composites have better mechanical properties than unreinforced aluminium composites; thus, the development of a reinforced aluminium feedstock is expected to have favourable properties that best suit the property requirements of relevant industrial applications.

REFERENCES

Abdel-Gawad, S. A., Osman, W. M., & Fekry, A. M. (2019). Characterization and corrosion behavior of anodized aluminium alloys for military industries applications in artificial seawater. *Surfaces and Interfaces*, *14*, 314–323.

Abdelrahman, E., Adel, N., & Galal A. (2020). Optimization of stir casting process parameters for producing MMC aluminum sacrificial anode incorporated with manganese dioxide concentrate powder. *International Journal of Engineering Research and Technology*, *13*(10), 2651–2659.

Adat, R., Kulkarni S., & Kulkarni, S. (2015). Manufacturing of particulate reinforced aluminum metal matrix composites using stir casting process. *International Journal of Current Engineering and Technology*, *5*(4), 2808–2812.

Adebayo, A., Adebisi, Abdul, M., & Mohammed, B. N. (2017). Influence of stirring speed on microstructure and wear morphology of SiCp-6061Al composite. *International Journal of Engineering Materials and Manufacture*, *1*, 21–26.

Ajagol, P., Anjan, B. N., Marigoudar, R. N., & Kumar, G. P. (2018, June). Effect of SiC reinforcement on microstructure and mechanical properties of aluminum metal matrix composite. In *IOP conference series: materials science and engineering* (Vol. 376, No. 1, p. 012057). IOP Publishing.

Alaneme, K. K., & Bamike, B. J. (2018). Characterization of mechanical and wear properties of aluminium based composites reinforced with quarry dust and silicon carbide. *Ain Shams Engineering Journal*, *9*(4), 2815–2821.

Amit, et al. (2020). Effect of B4C and waste porcelain ceramic particulate reinforcements on mechanical and tribological characteristics of high strength AA7075 based hybrid composite. *Journal of Material Research and Technology*, *9*(5), 9882–9894.

Amouri, K., Kazemi, S., Momeni, A., & Kazazi, M. (2016). Microstructure and mechanical properties of Al-nano/micro SiC composites produced by stir casting technique. *Materials Science and Engineering: A, 674*, 569–578.

Andersen, S. J., Marioara, C. D., Friis, J., Wenner, S., & Holmestad, R. (2018). Precipitates in aluminium alloys. *Advances in Physics: X, 3*(1), 1479984.

Arulraj, M., Palani, P. K., & Venkatesh, L. (2017). Optimization of process parameters in stir casting of hybrid metal matrix (LM25/SiC/B4C) composite. *Journal of Advances in Chemistry, 13*(11), ISSN 2321 - 807X 6038-6042| Page DOI: 10.24297/jac.v13i11.5774.

Ayogu, P. C., & Eze, F. I. (2019). Colorimetric determination of chromium in aluminium alloys by diphenylcarbazide method. *Open Journal of Chemistry, 5*(1), 9–12.

Bandil, K., Vashisth, H., Kumar, S., Verma, L., Jamwal, A., Kumar, D., … & Gupta, P. (2019). Microstructural, mechanical and corrosion behaviour of Al–Si alloy reinforced with SiC metal matrix composite. *Journal of Composite Materials, 53*(28–30), 4215–4223.

Brough, D., & Jouhara, H. (2020). The aluminium industry: A review on state-of-the-art technologies, environmental impacts and possibilities for waste heat recovery. *International Journal of Thermofluids, 1*, 100007.

Caetano, S. (2017). Overview of common aerospace aluminum alloys: 2024, 6061, and 7075, *1*, 1–5.

Chelladurai, S. J. S., & Arthanari, R. (2018). Effect of stir cast process parameters on wear behaviour of copper coated short steel fibers reinforced LM13 aluminium alloy composites. *Materials Research Express, 5*(6), 066550.

Daniel, B., & Jouhara, H. (2020). The aluminium industry: A review on state-of-the-art technologies, environmental impacts and possibilities for waste heat recovery. *International Journal of Thermofluids, 1*, 100007.

David Raja Selvam, J., Dinaharan, I., & Mashinini, P. M. (2017). High temperature sliding wear behavior of AA6061/fly ash aluminum matrix composites prepared using compocasting process. *Tribology-Materials, Surfaces & Interfaces, 11*(1), 39–46.

Dhanesh, S., Kumar, K. S., Fayiz, N. M., Yohannan, L., & Sujith, R. (2021). Recent developments in hybrid aluminium metal matrix composites: A review. *Materials Today: Proceedings, 45*, 1376–1381.

Doan, B. Q., Nguyen, D. T., Nguyen, M. N., Le, T. H., & Hao, T. M. (2021). A review on properties and casting technologies of aluminum alloy in the machinery manufacturing. *Journal of Mechanical Engineering Research and Developments, 44*(8), 204–217.

Elkilany, H. A., Shoeib, M. A., & Abdel-Salam, O. E. (2019). Influence of hard anodizing on the mechanical and corrosion properties of different aluminium alloys. *Metallography, Microstructure, and Analysis, 8*(6), 861–870.

Fayomi, O. S. I., & Akande, I. G. (2019). Corrosion mitigation of aluminium in 3.65% NaCl medium using hexamine. *Journal of Bio- and Tribo-Corrosion, 5*(1), 1–7.

Fayomi, O. S. I., Joseph, O. O., Akande, I. G., Ohiri, C. K., Enechi, K. O., & Udoye, N. E. (2019). Effect of CCBP doping on the multifunctional Al-0.5 Mg-15CCBP superalloy using liquid metallurgy process for advanced application. *Journal of Alloys and Compounds, 783*, 246–255.

Gazizov, M., & Kaibyshev, R. (2017). Precipitation structure and strengthening mechanisms in an Al-Cu-Mg-Ag alloy. *Materials Science and Engineering: A, 702*, 29–40.

Greer, A. (2016). Overview: Application of heterogeneous nucleation in grain-refining of metals. *The Journal of Chemical Physics, 145*(21), 211704.

Gu, J., Ding, J., Williams, S. W., Gu, H., Bai, J., Zhai, Y., & Ma, P. (2016). The strengthening effect of inter-layer cold working and post-deposition heat treatment on the additively manufactured Al63 Cu alloy. *Materials Science and Engineering: A, 651*, 18–26.

Haghayeghi, R., Zoqui, E. J., & Timelli, G. (2018). Enhanced refinement and modification via self-inoculation of Si phase in a hypereutectic aluminium alloy. *Journal of Materials Processing Technology*, *252*, 294–303.

Huda, A. A. et al. (2019). Mechanical and wear behavior of AA7075 aluminium matrix composites reinforced by Al_2O_3 nanoparticles. *Nanocomposites*, *5*(3), 67–73, doi: 10.1080/20550324.2019.1637576.

Kaya, H., & Aker, A. (2017). Effect of alloying elements and growth rates on microstructure and mechanical properties in the directionally solidified Al–Si–X alloys. *Journal of Alloys and Compounds*, *694*, 145–154.

Kulkarni, S. G., Menghani, J. V., & Lal, A. (2016). Investigation of mechanical properties of fly ash and Al_2O_3 reinforced A356 alloy matrix hybrid composite manufactured by improved stir casting, *Indian Journal of Engineering and Materials Sciences*, *23*(1), 27–36.

Kumar, A., Pal, K., & Mula, S. (2021). Effects of cryo-FSP on metallurgical and mechanical properties of stir cast Al7075–SiC nanocomposites. *Journal of Alloys and Compounds*, *852*, 156925.

Kumar, M. S., Begum, S. R., & Vasumathi, M. (2019). Influence of stir casting parameters on particle distribution in metal matrix composites using stir casting process. *Materials Research Express*, *6*(10), 1065d4.

Maneiah, D., Shunmugasundaram, M., Reddy, A. R., & Begum, Z. (2020). Optimization of machining parameters for surface roughness during abrasive water jet machining of aluminium/magnesium hybrid metal matrix composites. *Materials Today: Proceedings*, *27*, 1293–1298.

Mansor, M. R., Nurfaizey, A. H., Tamaldin, N., & Nordin, M. N. A. (2019). Natural fiber polymer composites: utilization in aerospace engineering. In *Biomass, biopolymer-based materials, and bioenergy* (pp. 203–224). Woodhead Publishing.

McKenna, R. C., & Norman, J. B. (2010). Spatial modelling of industrial heat loads and recovery potentials in the UK. *Energy Policy*, *38*(10), 5878–5891.

Mondiu, O. D., Johnson, O. A., Lateef, O. M., Adekunle, A. Y., Solomon, K. B., & Taiwo, O. R. (2017). Optimization of stir casting parameters to improve the hardness property of Al/RHA matrix composite. *EJERS, European Journal of Engineering Research and Science*, *2*(11), 5–12.

Nair, S., & Joshi, N. (2015). Preparation of Al 6061/ SiC metal matrix composite (MMC) using stir casting technique. *IJARIIE, ISSN(O)*, *1*(3), 2395–4396.

Nallusamy, S. (2016). A review on the effects of casting quality, microstructure and mechanical properties of cast Al-Si-0.3 Mg alloy. *International Journal of Performability Engineering*, *12*(2), 143.

Prakash, M., & Jaswin, M. (2015). Microstructural analysis of aluminum hybrid metal matrix composites developed using stir casting process, *International Journal of Advance Engineering*, *1*(3), 333–339.

Rakhmonov, J., Timelli, G., & Bonollo, F. (2016). the effect of transition elements on high-temperature mechanical properties of Al–Si foundry alloys: A review. *Advanced Engineering Materials*, *18*(7), 1096–1105.

Rana, R. S., Purohit, R., & Das, S. (2012). Reviews on the influences of alloying elements on the microstructure and mechanical properties of aluminum alloys and aluminum alloy composites. *International Journal of Scientific and Research Publications*, *2*(6), 1–7.

Rangrej, S., Mehta, V., Ayar, V., & Sutaria, M. (2021). Effects of stir casting process parameters on dispersion of reinforcement particles during preparation of metal composites. *Materials Today: Proceedings*, *43*, 471–475.

Reichardt, A., Shapiro, A. A., Otis, R., Dillon, R. P., Borgonia, J. P., McEnerney, B. W., ... & Beese, A. M. (2021). Advances in additive manufacturing of metal-based functionally graded materials. *International Materials Reviews, 66*(1), 1–29.

Riestra, M., Ghassemali, E., Bogdanoff, T., & Seifeddine, S. (2017). Interactive effects of grain refinement, eutectic modification and solidification rate on tensile properties of Al-10Si alloy. *Materials Science and Engineering: A, 703*, 270–279.

Saikrupa, C., Reddy, G. C. M., & Venkatesh, S. (2021, February). Aluminium metal matrix composites and effect of reinforcements – A review. In IOP Conference series: Materials science and engineering (Vol. 1057, No. 1, p. 012098). IOP Publishing.

Stemper, L., Mitas, B., Kremmer, T., Otterbach, S., Uggowitzer, P. J., & Pogatscher, S. (2019). Age-hardening of high pressure die casting AlMg alloys with Zn and combined Zn and Cu additions. *Materials and Design, 181*, 1–11.

Su, H., Gao, W., Feng, Z., & Lu, Z. (2012). Processing, microstructure and tensile properties of nano-sized Al_2O_3 particle reinforced aluminium matrix composites, *Materials and Design, 36*, 590–596.

Tamilanban, T., & Ravikumar, T. S. (2021). Influence of stirring speed on stir casting of SiC reinforced Al Mg Cu composite. *Materials Today: Proceedings, 45*, 5899–5902.

Tisza, M., & Czinege, I. (2018). Comparative study of the application of steels and aluminium in light-weight production of automotive parts. *International Journal of Light-Weight Materials and Manufacture, 1*(4), 229–238.

Varshney, D., & Kumar, K. (2021). Application and use of different aluminium alloys with respect to workability, strength and welding parameter optimization. *Ain Shams Engineering Journal, 12*(1), 1143–1152.

Vu, C., Wern, C., Kim, B. H., Kim, S. K., Choi, H. J., & Yi, S. (2019). Fatigue characteristic analysis of new ECO7175v1 extruded aluminium Alloy. *Journal of Aerospace Engineering, 32*(1), 1–6.

Xiao, P., Gao, Y., Xu, F., Yang, S., Li, B., Li, Y., ... & Zheng, Q. (2019). An investigation on grain refinement mechanism of TiB2 particulate reinforced AZ91 composites and its effect on mechanical properties. *Journal of Alloys and Compounds, 780*, 237–244.

Yan, P., Mao, W., Fan, J., & Wang, B. (2019). Simultaneous refinement of primary Si and modification of eutectic Si in A390 alloy assisting by Sr-modifier and serpentine pouring channel process. *Materials, 12*(19), 3109.

Yi, X., Lu, Y., & He, G. (2024). Aluminum demand and low carbon development scenarios for major countries by 2050. *Journal of Cleaner Production*, 143647.

Zamani, M. et al. (2017). The role of transition metal additions on the ambient and elevated temperature properties of Al-Si alloys. *Materials Science and Engineering: A, 693*, 42–50.

Zhang, W., Liaw, P. K., & Zhang, Y. (2018). Science and technology in high-entropy alloys. *Science China Materials, 61*(1), 2–22.

Zhao, D., Zhuang, Z., Cao, X., Zhang, C., Peng, Q., Chen, C., & Li, Y. (2020). Atomic site electro catalysts for water splitting, oxygen reduction and selective oxidation. *Chemical Society Reviews, 49*(7), 2215–2264.

8 Electrochemistry of Corrosion

8.1 INTRODUCTION

The study of electrochemistry considers the chemical processes involved in ionic transport-induced homogeneous charge transfer in an electrolyte and electric wire-induced heterogeneous charge transfer on electrode surfaces. Due to at least two half-cell reactions going in the opposite direction, this results in electroneutrality. The interfacial region between the electrolyte and electrode surface is where the electrochemical heterogeneity associated with oxidation and reduction reactions on the corresponding electrode surfaces exists (Jashari et al., 2022). In an electrochemical reaction thermodynamic tendency, for example, metal oxidation is determined by the change in free energy connected with that reaction as demonstrated in the Pourbaix equilibrium diagram. In a favourable thermodynamic electron transfer this process is initiated, and the occurrence of a spontaneous reaction is determined by thermodynamic reactions (i.e., without the need for an external source of energy or input of an external driving force). Homogenous reaction that involves the composition or generation of various ions species and electrons are used to produce great electrochemical processes at the interface of metal and solution leading to electrochemical reactions in an aqueous corrosion process. Corrosion will not occur in an aqueous environment if the metal desolation is not favoured by the thermodynamic criterion. Due to these reasons, noble metals like platinum and gold do not undergo corrosion in pure water (Ghanei et al., 2020).

8.1.1 Pourbaix Diagram

A Pourbaix diagram is an electrochemical map that illustrates ion borders as lines, similar to a phase diagram. It provides valuable insights into the stability of ions, oxides, and hydroxides, making it applicable in various fields such as corrosion studies, electrowinning, electroplating, hydrometallurgy, electrolysis, electrical cells, and water treatment (Eliaz, 2019). The diagram shows the oxidizing power of the electrochemical field and the pH levels indicating the acidity and alkalinity of species. While it aids in designing and analysing electrochemical systems by highlighting potential reactions and important areas, it does

DOI: 10.1201/9781003562078-8

not include information on corrosion rate, which is a crucial factor in kinetic investigations. To create a Pourbaix diagram for a metal M-H_2O electrochemical system, the main ion boundaries enclosing stable phases in an aqueous environment are depicted as lines.

8.2 KINETICS OF ACTIVATION POLARIZATION

When an electrochemical cell experiences the flow of electric current, it deviates from its equilibrium state, leading to a phenomenon called polarization. This polarization can be categorized into activation polarization and concentration polarization. Activation polarization occurs when the electrochemical cell deviates from equilibrium during current flow, resulting in oxidation reactions with a positive overpotential and reduction reactions with a negative overpotential. On the other hand, concentration polarization arises from mass transport due to a concentration gradient in the electrolyte solution, which is driven by diffusion. To accurately assess the corrosion rate of a metal (M) exposed to a corrosive environment, understanding electrochemical reaction kinetics is crucial. While thermodynamics can predict the potential for corrosion, it does not provide information on the speed of corrosion. The rate of electron movement at the electrode–electrolyte interface significantly influences the reaction rate. In equilibrium, the net reaction rate is zero. However, corrosion rates are primarily governed by electrochemical kinetics, while chemical kinetics dictate reaction rates (Roh, 2013). To evaluate the electrochemical kinetics of corrosion, the corrosion current density (i_{corr}) and corrosion potential (E_{corr}) are determined. These values help analyse the corrosion behaviour through a polarization curve (E vs. log i). By examining these characteristics, the corrosion rate (i_{corr}) can be calculated, often converted into the Faradaic corrosion rate (CR) with units of mm/yr, along with the polarization resistance (R_p).

8.2.1 Polarization Diagram

The phenomenon of polarization occurs when electrode reactions induce deviations from equilibrium as a result of the passage of electric current through an electrochemical cell, leading to changes in the potential of the working electrode (WE). This deviation from equilibrium creates an electric potential difference between the polarized electrode and the unpolarized electrode, known as overpotential (Mahmoud et al., 2021).

8.2.2 Tafel Extrapolation

Multiple studies have investigated the corrosion rate and resistance of metals in various media (Jain et al., 2020; Aslam et al., 2022). Typically, the corrosion rate equation (Equation 8.1) is used to compare the corrosion resistance of protected and unprotected/as-received metals to determine if sacrificial protection is effective. The addition of inhibiting materials or alloying elements with anodic properties often reduces the dispersion of corrosion products and contaminant ions into the metal's core and surface, thereby enhancing corrosion resistance (Popova & Prosek, 2022). These inhibiting substances or protective alloying elements have also proven effective in preventing cathodic corrosion in chloride and sulphide-containing corrosive environments. To evaluate the corrosion resistance of materials like steel, copper, and aluminium used in marine and industrial settings, techniques such as weight loss or gravimetric experiments and potentiodynamic polarization tests have been widely employed (Elewa et al., 2019). Potentiodynamic polarization tests are preferred due to their accuracy and efficiency. However, weight loss experiments are conducted when the cost of purchasing an electrochemical cell is a concern. Although weight loss trials are cost-effective and less complex, they may take several days to complete, compromising their precision and accuracy (Jiang et al., 2017; Xia et al., 2019). The corrosion rate can be calculated using various methods outlined in Equations (8.1)–(8.3).

$$cpy = \frac{87.6W}{DAT} \tag{8.1}$$

where corrosion penetration per year is in mm/yr, D is the specimen density in g/cm^3, W is the weight loss in mg, the specimen area A is measured in sq., and the exposure time T is measured in hours. Weight loss data are taken from the corrosion experiment (Kotes et al., 2018). Equation (8.2) shows similar expressions.

$$\frac{\mu\text{m}}{\text{yr}} = 87{,}600\frac{W}{DAT} \tag{8.2}$$

where W represents the weight loss in milligrams, D density in g/cm^3, A is an area in cm^2, and T is the time of exposure in hours. The corrosion rate is the information extracted from potentiodynamic polarization scan. The current density can be expressed for reactions that are principally activation controlled (i.e., mass transfer controlled or change transfer-controlled reactions occurring at a lesser rate compared to the limiting rate), the function of overpotential (η) is also expressed in current density, as shown in Equations (8.3) and (8.4).

$$\eta = \beta \log i / i0 \tag{8.3}$$

where

$$\eta = E_{\text{applied}} - E_{\text{opencircuit}} \tag{8.4}$$

The expression in Equation (8.3) is identified as the Tafel equation, where I is the current density, β is the Tafel slope, and i_0 is the current exchange density (reversible potential at the reaction rate of the specific reaction). Therefore, the linear regions of the polarization curve can be used to obtain the Tafel slope for the anodic and cathodic reactions occurring at an open circuit (He et al., 2022).

8.2.3 Impedance Spectroscopy

Determining fundamental variables like polarization resistance (R_p) and capacitance is necessary to determine the polarization impedance of an electrode (Cdl). The working electrode's impedance is then measured and used as a helpful kinetic reference. It is important to understand that when an electrode is submerged in an electrolyte, metal and electrolyte ions combine to form an ionic layer that causes a charge distribution at the interfacial region with special properties related to the half-cell inner potential and the electric double-layer impedance. This suggests that a current flow perturbation is present at the electrode–electrolyte interface. The quantity of the current passing through it and the sort of ionic reaction that may occur inside the thickness of the contact, theoretically, upset this interface (Law et al., 2000).

8.3 CORROSIVITY AND PASSIVITY

Passivity is a phenomenon in which the surface of metallic objects and alloys is characterized according to the formation of a highly protective thin invisible film with unique properties when interacting with a corrosive environment, making the metal or alloy noble (Tomashov, 2012). The formation of these films results in a metal or alloy surface becoming resistant (passive) to corrosive attack. The passivity phenomenon underlies the successful use of many active metals and alloys in industrial applications. In major localized failure, corrosion occurs due to passivity breakdown, which is responsible for intergranular, crevice, stress corrosion cracking, and pitting. Likewise, studies have shown that iron's passivity can be attained using anodic polarization. A metal active in the EMF series is passive when its electrochemical characteristics become that of a noble metal (Tailleart et al., 2012). In the EMF series, noble potential possesses a low corrosion rate. Examples of alloys and metals exhibiting passivity include chromium, nickel, steels, iron in oxidizing environments, aluminium, titanium, and many others (Scully et al., 2020).

8.3.1 Behaviour of Active and Passive Nature of Metals

A polarization curve, which is a plot of voltage vs. current density and is recorded under dynamic or steady-state settings, is used to identify systems that exhibit passivity. By tracking the variations in current density with reference to voltage or detecting variations in voltage with current density, one can obtain the steady-state polarization curve. Anodic partial current densities on a metal or alloy I vs. E exhibit an active zone [A-B], passivity, and trans passive behaviour. The active charge transfer controlled limit of the metal dissolution rate is indicated by the E–I data from A–B. I exhibits a critical current density, i_{crit}. The passive current density, i_{pass}, is given at D for the passive range, which is given from E to F. While EP2 denotes an activation or reactivation potential, EP1 stands for the main passive potential. On semi-conducting or electronically conducting oxide-covered metals, FGH reflects the oxygen evolution reaction rate, whereas trans passivity is observed and is indicated by the dotted line on select metals (Scully & Lutton, 2018). Metal will be regarded as passive if, upon advancing its potential toward higher positive or anodic magnitudes, there is a decrease in the rate of anodic dissolution when a potential is reached compared to the value observed that is less at the lesser anodically potential (Fayomi et al., 2024). The result of the buildup of passive film is the reduction in anodic dissolution rate; despite that, the thermodynamic tendency for metal dissolution is higher. For the process of film formation, two important mechanisms have been proposed; the first mechanism is known as the dissolution–precipitation mechanism (Maher et al., 2022).

8.4 EXPERIMENTAL DATA

8.4.1 POTENTIOSTATIC POLARIZATION

The potentiostatic polarization method involves measuring the current at a specific time while incrementally increasing the electric potential supplied by a potentiostat from a predetermined level. This technique, commonly used for anodic polarization, can also be applied to cathodic studies, as demonstrated by the potentiostatic polarization curve for AISI 1080 carbon steel in de-aerated 1 N sulphuric acid (H_2SO_4) solution at room temperature (Jacobs et al., 2019). The polarization map clearly separates the anodic and cathodic regions along with their respective kinetic parameters near the corrosion point. The electrochemical property E_{corr} 0:51 VSCE can be observed in this diagram, depending on the experimental conditions. The potentiostatic polarization method allows for the characterization of a metal's anodic and cathodic behaviour in an electrolyte, similar to galvanostatic and potentiodynamic methods. In general, potentiostatic polarization tests are valuable for identifying corrosion products, assessing corrosion reactions on a metal surface, and selecting suitable alloys.

8.4.2 LINEAR POLARIZATION CURVE TECHNIQUE

Linear polarization resistance (LPR) is a method used in monitoring the rate of corrosion in a short period. Corrosion current (I_{Corr}) is found using the application from the Stern-Geary equation, as shown in Equations (8.5)–(8.7), where B is a constant determined by Tafel slopes, and R_p is the polarization resistance.

$$I_{CORR} = R_p \tag{8.5}$$

$$B \frac{(\beta a x \beta c)}{2.3(\beta a + \beta c)} \tag{8.6}$$

$$R_p = \frac{\Delta E}{\Delta I} \tag{8.7}$$

ΔE represents the potential amplitude shift away from corrosion potential, and ΔI represents a change in the measured current. A range of ±5 to ±10 mV was recommended by Kelly (2022) to use usually although (Trethewey & Chamberlain, 1995) used ±30 mV. The linear polarization technique is used in the field and laboratory tests due to its relatively simplicity and it does not obstruct the corrosion system.

On the other hand, the main drawback of this method is that the B (constant) has to be assumed, which significantly influences the obtained results. The active B value in the Stern-Geary equation was proposed by González et al. (2007) to be 26 mV while the passive state is 52 mV. Stefanoni, Angst, & Elsener (2019) also claimed that using a B value of 26 mV to determine the rate of metal corrosion has an error factor of two. A polarization curve can provide evidence of whether or not a material is active, passive, or active-passive. Hence, the potentiostatic testing method can give reproducible results to determine the passivity of a metal and the corrosivity of an electrolyte.

8.4.3 Potentiodynamic Polarization

This method makes use of a current logarithmic converter and a programmable potentiostat to provide an ever-increasing potential (E) at a chosen potential scan rate ($dE = dt$) (Parangusan et al., 2021). The electrochemical behaviour of a working electrode can then be described and its kinetic characteristics extracted using a potentiodynamic polarization diagram. A polarization diagram created using potentiostatic and potentiodynamic polarization methods is equivalent, incidentally. A potentiodynamic polarization is frequently used to evaluate a metal's anodic behaviour, but the cathodic curve can also be included using an applied potential range. Utilizing a three-electrode cell, the potentiodynamic polarization experiment (PPE) has been used to examine the corrosion characteristics of metallic materials. The metallic material serves as the working electrode (WE) in a three-electrode cell, saturated calomel (SCE) serves as the reference electrode, and platinum or graphite is used as the counter electrode (CE). Frequently, this experiment is conducted in accordance with ASTM G5-94 guidelines (Tabelin et al., 2017). Some of the corrosion resistance indication metrics that can be obtained from the PPE are the open-circuit potential (OCP) and the polarization resistance (Pr). The steady-state potential of metallic materials is known as the OCP. To determine whether the metallic material is positively or negatively polarized in a test media, the results of corrosion potential and OCP can be compared. For example, a metal is considered to be positively polarized if its OCP value is less positive than the equivalent E_{corr} values, whereas a metal is said to be negatively polarized if its OCP value is higher. To investigate the performance of the metal in the test media, a further plot of the OCP vs. exposure duration graph might be made. On the other hand, greater levels of Pr show that the metal is effectively protected against corrosion; hence alloyed metals frequently have greater Pr than the as-received. A conceptual anodic polarization curve was observed from the anodic scan (Virtanen, 2011).

In metal dissolution, passivity breakdown is responsible for localized corrosion failures and intergranular corrosion. In general, passive films can be damaged or removed as protective layers by the following processes: Anodic breakdown passivity is also known as electrochemical de-passivation, chemical de-passivation, chemical film dissolution, undermining of porous films, reductive dissolution of oxides, and mechanical stress. A critical potential E_{corr} takes place that is equalled or is surpassed for passivity breakdown occurrence, while below its metal remains passive. This potential is known as pitting or breakdown potential, measured by different techniques, including potentiostatic methods, galvanostatic methods, and mechanical surface scratching (Parfenov et al., 2020).

8.4.4 Activation Polarization

Activation polarization is related to potential differences beyond the required equilibrium necessary to generate currents, which invariably depends on the activation energy of combined oxidation and reduction reactions. It can be referred to as the energy in the minimum quantity needed to energize atoms or molecules required for electrons to move from electrodes into a substance whose chemical constituents are being identified and measured (Perez, 2016). Activation polarization is referred to as electron transfer. Activation polarization is among the factors within the electrochemical reactions inherent to kinetics. Hydrogen gas evolution is a typical example. These reactions are a simple type of reaction; the rate of hydrogen ions' transformation to hydrogen gas controls numerous factors like the rate of transfer of electrons into ions of hydrogen. Different metals are used to categorize by electron mobility rates. Consequently, the hydrogen formation rates of various metals also vary significantly. Thc formation of hydrogen and cathode surfaces is used to develop these categories of polarization (Tao et al., 2020).

8.4.5 Resistance Polarization

Resistance polarization is also known as ohmic polarization. Current is transported from the anode to the cathode by ions in the electrolyte and in the metallic path from the anode to the cathode (Omoegun et al., 2023). Because of the normal high conductivity of metals, almost no resistance is offered to the current flow in the metallic path. However, resistance can be encountered if the distance between the anode and the cathode is appreciable (Torkaman & Karimi, 2015). At point R1 there was an intersection of the cathodic and the anodic curve with seawater as the electrolyte, where the potentials of the anode and the cathode are polarized at the same value. The resistivity of the solution is high at a potential drop (IR) resulting from the flow of the current in the resistive solution and the potential of the anode and cathode differ slightly in the case of seawater. The reason is that the potential drop (IR) diminishes the driving force of activation polarization. The anodic and cathodic reactions are not polarized to the same potentials. This situation is represented by R2 and further increasing the electrolyte resistance, the magnitude of the ohmic drop would further increase, as shown by R3. With an increasing resistance offered by the electrolytes, the magnitude of the corrosion current decreases as shown by R1, R2, and R3. Incidentally, increasing the solution resistance offers good methods of controlling corrosion, since decreasing corrosion current means decreasing corrosion rate (Teijido et al., 2022).

8.5 WEIGHT LOSS ANALYSIS

A gravimetric experiment involves submerging the metal in the test fluid. Using thread that is tightly fastened to a clamp, the metal is immersed in a solution. To avoid corrosion or solution reaction, the thread is frequently selected with caution.

The corrosive medium is often made using distilled water and a measured amount of alkaline solution, acidic solution, or salt. The metal is scraped and cleansed to remove corrosion flakes before being submerged in the test solution for a specified amount of time. The weight reduction (*W*) is then computed by subtracting the original weight from the new weight of the metal (weight before immersion in the test medium). Therefore, Equation (8.8) indicates how the corrosion rate (CR) can be calculated.

$$\text{CR} = \frac{87.6W}{TAD} \tag{8.8}$$

where *D* is the density of the metal, *T* is the time of exposure in the medium, and *A* is the metal exposed surface area in the medium and the unit of corrosion calculated is $g{\cdot}m^{-2}{\cdot}d^{-1}$. The corrosion protection efficiency of alloys or devices can be obtained from Equation (8.9).

$$\text{EC} = \frac{\text{Wo} - \text{Wi}}{\text{Wo}} \times 100 \tag{8.9}$$

where Wo is referred to as the weight loss of the control sample or as received samples and Wi is the weight loss in the presence of corrosion protection, respectively (He et al., 2021).

8.6 VOLTAMMETRY

8.6.1 Linear Sweep Voltammetry (LSV)

Linear sweep voltammetry (LSV), one of many electrochemical techniques discovered elsewhere (Labib, Sargent & Kelley, 2016), is briefly discussed from an analytical standpoint. This voltammetric technique measures the current response by applying a time-dependent linear potential sweep to an electrode. A current–potential plot of the experimental data produced under circumstances that allow polarization of a working electrode is shown. Since it links the potential (*E*) and current (*I*) of an electrochemical cell, cyclic voltammetry is essentially a potentiodynamic electrochemical technique. Schematic plots for the reversible reaction of different scan rates (van Deelen et al., 2019).

REFERENCES

Arakere, N. K. (2016). Gigacycle rolling contact fatigue of bearing steels: A review. *International Journal of Fatigue*, *93*, 238–249.

Aslam, J., Verma, C., & Hussain, C. M. (Eds.). (2023). *Electrochemical and analytical techniques for sustainable corrosion monitoring: Advances, challenges and opportunities 59–75.*

Elewa, R. E., Afolalu, S. A., & Fayomi, O. S. I. (2019, December). Overview production process and properties of galvanized roofing sheets. In *Journal of Physics: Conference Series* (Vol. 1378, No. 2, p. 022069). IOP Publishing.

Eliaz, N. (2019). Corrosion of metallic biomaterials: A review. *Materials*, *12*(3), 407.

Fayomi, O. S. I., Atiba, J. O., & Dauda, K. T. (2024). Electrochemistry and inhibitory evaluation of Musa paradisiaca particulates on AA6063 alloy for improved service life. *Hybrid Advances*, *6*(June), 100234.

Ghanei, A., Eskandari-Naddaf, H., Ozbakkaloglu, T., & Davoodi, A. (2020). Electrochemical and statistical analyses of the combined effect of air-entraining admixture and micro-silica on corrosion of reinforced concrete. *Construction and Building Materials*, *262*, 120768.

He, S., Wang, J., Zhang, J., & Xu, J. (2021). Intermittent versus continuous energy restriction for weight loss and metabolic improvement: A meta-analysis and systematic review. *Obesity*, *29*(1), 108–115.

He, W., Zhang, J., Dieckhöfer, S., Varhade, S., Brix, A. C., Lielpetere, A., ... & Schuhmann, W. (2022). Splicing the active phases of copper/cobalt-based catalysts achieves high-rate tandem electroreduction of nitrate to ammonia. *Nature Communications*, *13*(1), 1129.

Jacobs, M. H., Schmid-Fetzer, R., & van den Berg, A. P. (2019). Thermophysical properties and phase diagrams in the system MgO–SiO 2–FeO at upper mantle and transition zone conditions derived from a multiple-Einstein method. *Physics and Chemistry of Minerals*, *46*, 513–534.

Jain, P., Patidar, B., & Bhawsar, J. (2020). Potential of nanoparticles as a corrosion inhibitor: a review. *Journal of Bio- and Tribo-Corrosion*, *6*, 1–12.

Jashari, G., Švancara, I., & Sýs, M. (2022). Characterisation of carbon paste electrodes bulk-modified with surfactants for measurements in nonaqueous media. *Electrochimica Acta*, *410*, 140047.

Jiang, T., Ren, L., Jia, Z. G., Li, D. S., & Li, H. N. (2017). Pipeline internal corrosion monitoring based on distributed strain measurement technique. *Structural Control and Health Monitoring*, *24*(11), e2016.

Kotes, P., Strieska, M., & Brodnan, M. (2018, July). Sensitive analysis of calculation of corrosion rate according to standard approach. In *IOP conference series: Materials science and engineering* (Vol. 385, No. 1, p. 012031). IOP Publishing.

Law, D. W., Millard, S. G., & Bungey, J. H. (2000). Linear polarisation resistance measurements using a potentiostatically controlled guard ring. *NDT & E International*, *33*(1), 15–21.

Maher, M., Iraola-Arregui, I., Youcef, H. B., Rhouta, B., & Trabadelo, V. (2022). The synergistic effect of wear-corrosion in stainless steels: A review. *Materials Today: Proceedings*, *51*, 1975–1990.

Mahmoud, W. M., Elfiky, D., Robaa, S. M., Elnawawy, M. S., & Yousef, S. M. (2021). Effect of atomic oxygen on LEO CubeSat. *International Journal of Aeronautical and Space Sciences*, *22*, 726–733.

Omoegun, O. G., Fayomi, O. S. I., & Atiba, J. O. (2023). Investigation of the corrosive behavior and adsorption parameters of copper in a cowbone ash inhibited alkaline environment. *Journal of Bio- and Tribo-Corrosion*, *9*(4), 75.

Parangusan, H., Bhadra, J., & Al-Thani, N. (2021). A review of passivity breakdown on metal surfaces: Influence of chloride-and sulfide-ion concentrations, temperature, and pH. *Emergent Materials*, *4*(5), 1187–1203.

Parfenov, E. V., Kulyasova, O. B., Mukaeva, V. R., Mingo, B., Farrakhov, R. G., Cherneikina, Y. V., ... & Valiev, R. Z. (2020). Influence of ultra-fine grain structure on corrosion behaviour of biodegradable Mg-1Ca alloy. *Corrosion Science*, *163*, 108303.

Perez, N. (2016). Kinetics of activation polarization. In *Electrochemistry and corrosion science* (pp. 101–150). Springer, Cham.

Popova, K., & Prošek, T. (2022). Corrosion monitoring in atmospheric conditions: A review. *Metals*, *12*(2), 171.

Roh, H. S. (2013). Kinetics of polarization mechanisms. *Journal of the Electrochemical Society*, *160*(9), H519.

Scully, J. R., Inman, S. B., Gerard, A. Y., Taylor, C. D., Windl, W., Schreiber, D. K., ... & Frankel, G. S. (2020). Controlling the corrosion resistance of multi-principal element alloys. *Scripta Materialia*, *188*, 96–101.

Scully, J. R., & Lutton, K. (2018). Polarization behavior of active passive metals and alloys. *Encyclopedia of Interfacial Chemistry: Surface Science and Electrochemistry*, *1*, 439–447.

Stefanoni, M., Angst, U. M., & Elsener, B. (2019). Kinetics of electrochemical dissolution of metals in porous media. *Nature Materials*, *18*(9), 942–947.

Tabelin, C. B., Veerawattananun, S., Ito, M., Hiroyoshi, N., & Igarashi, T. (2017). Pyrite oxidation in the presence of hematite and alumina: II. Effects on the cathodic and anodic half-cell reactions. *Science of the Total Environment*, *581*, 126–135.

Tailleart, N. R., Huang, R., Aburada, T., Horton, D. J., & Scully, J. R. (2012). Effect of thermally induced relaxation on passivity and corrosion of an amorphous Al–Co–Ce alloy. *Corrosion Science*, *59*, 238–248.

Tao, H., Zhou, C., Hong, Y., Zheng, Y., Zhang, K., Zheng, J., & Zhang, L. (2020). Influence of warm predeformation temperature on the corrosion property of type 304 austenitic stainless steel. *Journal of Materials Engineering and Performance*, *29*, 4515–4528.

Teijido, R., Ruiz-Rubio, L., Echaide, A. G., Vilas-Vilela, J. L., Lanceros-Mendez, S., & Zhang, Q. (2022). State of the art and current trends on layered inorganic-polymer nanocomposite coatings for anticorrosion and multi-functional applications. *Progress in Organic Coatings*, *163*, 106684.

Tomashov, N. D. (2012). *Passivity and protection of metals against corrosion*. Springer Science & Business Media.

Torkaman, H., & Karimi, F. (2015). Measurement variations of insulation resistance/polarization index during utilizing time in HV electrical machines–A survey. *Measurement*, *59*, 21–29.

Trethewey, K. R. (1995). Corrosion for science and engineering. *British Library Cataloguing in Publication Data, 99, 50–57.*

van Deelen, T. W., Hernández Mejía, C., & de Jong, K. P. (2019). Control of metal-support interactions in heterogeneous catalysts to enhance activity and selectivity. *Nature Catalysis*, *2*(11), 955–970.

Virtanen, S. (2011). Biodegradable Mg and Mg alloys: Corrosion and biocompatibility. *Materials Science and Engineering: B*, *176*(20), 1600–1608.

Xia, D. H., Song, S., Qin, Z., Hu, W., & Behnamian, Y. (2019). Electrochemical probes and sensors designed for time-dependent atmospheric corrosion monitoring: fundamentals, progress, and challenges. *Journal of the Electrochemical Society*, *167*(3), 037513.

9 High Entropy Alloys
The Study on Aluminium Cooking Ware

9.1 INTRODUCTION

Aluminium is one of the most used metals in the manufacturing industry due to its unique properties. However, the detrimental effects of aluminium (Al) degradation when it is used as a cookware have become a subject of global significance due to its harmful impact on human health and ecosystem (Belkhiri et al., 2017). Al alloys are also prone to acidifying problems. Aluminium may build up in plants and cause serious health issues for animals that eat these plants. Another harmful ecological effect of aluminium is that its ions can react with phosphates and reduce the availability of phosphates for water organisms (Goswami et al., 2019). There are also strong indications that aluminium can destroy the roots of trees situated in groundwater because aluminium of high concentration in the soil could be toxic (Ayeni et al., 2014; Elisa et al., 2016). Aluminium is extensively used in the industry for household utensils, transport, and packaging materials. However, some of the major challenges facing the use of aluminium alloy as cookware are its dissolution, wear or leaching into food during cooking. The longer food is cooked or stored in aluminium or aluminium alloy, the greater the amount that diffuses into food (Weidenhamer et al., 2017). Cooking with aluminium cookware could increase the aluminium content of foods, which could lead to severe health issues such as dementia, damage to the central nervous system, memory loss, and severe trembling (Al Juhaiman, 2016).

More so, minute acidic content in some foods such as tomatoes and essential foods like cereals may cause more aluminium leaching rate than usual to enter the food and can cause pitting on the pot's surface, especially if fluoride is present (Jabeen et al., 2016; Mol & Ulusoy, 2020). Therefore, aluminium cookware was banned in most European countries because all vegetables cooked in aluminium produce hydroxide poison, which can result in stomach and gastrointestinal troubles, such as stomach ulcers, colitis, and Alzheimer's disease (Bondy, 2016; Kandimalla et al., 2016). However, for economic reasons, it has been difficult to prevent people from using aluminium kitchen utensils in most parts of the world in spite of many reported cases of aluminium toxicity and health impact (Klotz et al., 2017; Klein, 2019). Thus, numerous aluminium cookware is still being produced and used in developing countries (Akbar & Nurpita, 2019; Gupta

DOI: 10.1201/9781003562078-9

et al., 2019). Moreover, recent studies also revealed that food prepared in aluminium cookware can be contaminated with Al ions which are invariably ingested into the body. This could be hazardous and consequently result in a series of health issues and the introduction of pathologies into the digestive system after prolonged usage (Ghasemidehkordi et al., 2018; Gatti & Montanari, 2018). World Health Organization (WHO) reported a day-by-day consumption of aluminium of about 9 mg by young and matured females, while for young and matured males a daily intake of 12–14 mg of aluminium was reported (Odularu et al., 2013). These quantities of consumption could be injurious to health.

9.2 ALUMINIUM COOKWARE CHALLENGES AND IMPACTS

Aluminium utensils are commonly used because they are readily available, cheap, lightweight, and exhibit good thermal conductivity. However, aluminium is known to neutralize the minerals and vitamins content of food (Ilich & Kerstetter, 2000). Phosphorous uptake by the intestine is subdued by aluminium due to its ability to bind inorganic phosphorous compounds, which could result in poor bone mineralization or bone softening (Stahl et al., 2018). Studies have shown that continuous exposure or consumption of aluminium dust builds up cough and presents an unnatural chest in X-rays. Besides, individuals with kidney problems may also have complexity in getting rid of surplus aluminium from their bodies (Nathan et al., 2019). Accumulation of aluminium in the body also poses risks for brain and bone disorders. Although there is no scientific evidence that has suggested that aluminium causes cancer, however, a high concentration of aluminium increases the risk of breast cancer (Jennrich & Schulte-Uebbing, 2016).

The gradual dissolution of aluminium cookware could result in pitting corrosion and loss of mechanical properties like hardness, strength, wear resistance, impact strength, thermal resistance, and at times ultimate failure. However, the vulnerability of aluminium to pitting depends on a number of factors, such as pH, the concentration of chloride ions, dissolved oxygen in the deterioration environment and surface conditions (Huang et al., 2016; Grgur & Marunkic, 2018; Soares et al., 2019). In addition, aluminium and its alloy can themselves contribute to pitting due to preferential etching. Pitting corrosion of aluminium often takes place as local damage to the surface. It frequently results in aesthetic damage at initiation, rather than functional damage. Nevertheless, a single pit in a crucial area can cause great damage (Hu et al., 2016). Deformities in the points where holes are bored for the insertion of screws can instigate pitting and greatly reduce the accuracy of holes' dimensions. Pitting corrosion can also assist in the initiation of stress corrosion cracking of aluminium cookware, and failure becomes eminent within a short period of usage (Goebel et al., 2016; Mai & Soghrati, 2017).

The pitting corrosion of metals is more conventionally described as an autocatalytic process. Pitting of metal results in localized acidity sustained by the spatial partition of anodic and cathodic half-reactions, which creates a gradient in potential and electro-transfer of destructive anions into the pit. For instance, the presence of metal in an oxygenated sodium chloride electrolyte produces a situation where the metal surface behaves as a cathode and pit as the anode. The production of ions of positive metal in the pit creates a local excess of positive charge, which pulls towards itself the negative chloride ions from the electrolyte to create neutrality of charge. Inside the pit is a high concentration of chloride molecules that react with water to produce hydrogen chloride gas, hydroxide of the metal and hydrogen ion, speeding up the corrosion process (Martinez-Salazar et al., 2020). More so, the concentration of oxygen inside the pit is basically zero and outside the pit on the metal surface, all the cathodic oxygen reactions take place. Although the corrosion initiation is autocatalytic in nature, its proliferation is not. Pitting corrosion is extremely dangerous, as it leads to minute loss of material with small visible surface defects, while it destroys the deep structure of the metal. The pits on the outside are often covered by corrosion products (Li et al., 2018; Melchers, 2020).

Aluminium cookware could as well be attacked by high-temperature corrosion and microbial corrosion, which could deteriorate the metal as a result of regular heating and incessant exposure to microorganisms, respectively. In high-temperature corrosion, a non-galvanic form of corrosion takes place when a metal is subjected or exposed to a hot environment containing sulphur, oxygen, or other compounds that have the tendency to oxidize or assist the oxidation of the concerned material (Goyal et al., 2017; Mannava et al., 2019). On the other hand, microbial corrosion also referred to as microbiologically prone corrosion could deteriorate metals with or without oxygen. Sulphate-reducing agents are energetic without oxygen (anaerobic). They generate hydrogen sulphide, causing stress cracking. In the presence of oxygen, some microbes may oxidize directly to oxides and hydroxides. However, others may oxidize sulphur to sulphuric acid, leading to biogenic sulphide corrosion (Al-Tamimi & Mehdi, 2017; Guan et al., 2017; Gu et al., 2019). Microbial corrosion mostly occurs in aluminium cookware that is left dirty for a long period or improperly washed.

Corrosion of metals is a universal and costly challenge for a number of industries. Understanding and reducing the cost of corrosion remain the key interest for most countries, corrosion specialists, and relevant asset owners (Hou et al., 2017). A precise knowledge of corrosion cost has been of immense interest to corrosion engineers and scientists for many centuries, and the detailed assessment of corrosion costs amongst different companies or industrial sectors presents the prospect of recognizing the common setbacks and top practices in corrosion reduction. Quantifying the overwhelming cost of corrosion is also a crucial measure to increase consciousness of the significance and enormity of corrosion damage. This will help to improve capacities for mitigating corrosion risks. The cost of corrosion on the economy could roughly be determined directly from the utilization and maintenance of corrosion-prevention technologies. Those technologies are protective coatings and inhibitors, cathodic and anodic protection. Others are the use of corrosion-resistant materials, and corrosion monitoring and inspection tools, or indirectly from productivity losses, pollution of the environment, a reward for casualties, and any other circuitous cost within that industry. Although the indirect cost of corrosion is complex to quantify the direct cost, it is considered computable via the combination of suitable methods such as extrapolation, statistics, and questionnaires, and by applying the information in both corrosion and economics (Ossai et al., 2015; Miroshnikova & Kuchugin, 2015).

More so, in Nigeria, the entire cost and the total annualized value of corrosion from 2013 to 2015 were approximately valued at ₦166,955,641 and ₦93,791,024, respectively (Orisanmi et al., 2017). The study on the state of corrosion and control policy in China estimated that corrosion cost in China was roughly 2127.8 billion RMB (about 310 billion U.S. dollars), which represents virtually 3.34% of the GDP (Hou et al., 2017). Based on the outcome of the survey, corrosion remains a principal and major issue that requires urgent and sustainable tactical approaches to combating it. The research on the significant impact of corrosion on metallic structures on the economy of the United States also estimated that the overall economic implication of corrosion and its control applications was worth 276 billion U.S. dollars annually. More so, the cost of corrosion in the U.S. military was estimated at 10–20 billion U.S. dollars. However, the application of best practice maintenance procedures was recommended for significant savings over time (Thompson et al., 2007; Iqbal et al., 2017). Studies have also shown that the corrosion cost incurred by most nations is generally estimated as 1%–5% of their GNP, and the global cost of corrosion was valued at 2.5 trillion U.S. dollars, representing 3.4% of the global product (Koch et al., 2016). This estimation was obtained by evaluating the accessible data from different regions of the world, and to provide efficient and sustainable corrosion control measures.

9.3 ALUMINIUM CORROSION AND STRUCTURAL IMPACT

Pitting corrosion, stress corrosion cracking, and leaching/wear are the most regularly encountered failures by aluminium components, especially cookware and baking tray materials (Jekle et al., 2016). According to Al Juhaiman (2016) and Weidenhamer et al. (2017), "leaching and pitting of aluminium cookware are a fundamental concern to the academic and the entire globe due to the impact on health and loss in mechanical properties over time".

Jabeen et al. (2016) affirmed that salty or acidic foods could increase the leaching rate of aluminium cookware or pots more than usual when they come in contact with the food. According to the authors "souring agents such as vinegar, lemon and tomato are now more included in cooking, increasing aluminium leaching". Aluminium possesses passive characteristics in aqueous solution because of the defensive compact Al_2O_3 film on its exterior or surface. However, these defensive compacts become increasingly soluble in alkaline and acidic mediums and erode within a period of exposure. A conclusion was reached that the entire specimens exposed to aluminium exhibited increased aluminium concentration, with foodstuff containing citric acid leaching aluminium at 292.25 mg/kg. The migration of aluminium into food was generally attributed possibly to food pH, raw materials chemical composition, cooking temperature, and availability of organic salts and acids. Exceeding the acceptable daily consumption of aluminium was adjudged to have several adverse effects on health. Based on the recommendation of the authors, aluminium cookware should not be heated dry because it might cause leaching of the metal. Furthermore, the use of aluminium cookware for a long period, especially for cooking salty and acidic food was discouraged due to leaching.

Investigation to Al Juhaiman (2010) "the erosion of aluminium in aqueous solutions could be explained knowing the acidic, saline or alkaline chemical reaction taking place on the surface of the aluminium cookware". The study carried out by the author revealed that cooking with aluminium cookware does not have a health-damaging effect on some regions of the world, but with so much acidic substance in the environment aluminium leaching might be enhanced. This could increase the aluminium absorption capacity of food and facilitate the injection of aluminium into the body. Leaching from aluminium cookware was also found to depend largely on the composition of the solution, concentration, and composition of aluminium (alloying element). In conclusion, foods containing lactic acid and amino acids were discovered to have a high vulnerability to react with A3+ to form an Al complex.

Martínez et al. (2015) affirmed that the wrong choice of material can result in an inadequate life cycle assessment of cookware. The study on the analysis of corrosion in different induction cookware materials was carried out by the authors in a salt spray chamber to evaluate and compare the corrosion properties of aluminium, stainless steel, and enamelled cast iron cookware to determine the material best for induction cooker in terms of corrosion. The cookware was positioned in a suspended position for 24 hours at a constant temperature of 30°C. It was observed that cookware made from stainless steel and enamelled cast iron exhibited high corrosion resistance compared to aluminium cookware, inside and outside. The outcome of this research was ascribed probably to the wrong material choice, which results from the use of recycled aluminium material as an alternative to using aluminium alloy appropriate for food.

Literature from Liang et al. (2008) "ageing time and temperature have a considerable effect on the mechanical properties, microstructure and intergranular corrosion of aluminium and its alloys". The author reported that the addition of special elements such as scandium (Sc) can enhance the properties of age-hardenable aluminium alloys. An investigation was carried out on AlCuLiZr and AlCuLiZrSc alloys at the maximum ageing temperature and time of 190°C and 72 hours, respectively. Generally, AlCuLiZrSc alloy exhibited better mechanical and corrosion resistance properties than AlCuLiZr alloy. However, the corrosion rate of AlCuLiZrSc alloy was found to increase with an increase in ageing time at a holding temperature of 160°C. The rate of corrosion increased from 1.0781 to 1.1193 and finally to 2.0898 mm y^{-1} for 24, 40, and 50 hours of ageing time, respectively. The hardness value was higher at 160°C/40 hours than ageing at 130°C/40 hours and 190°C/40 hours, but lower than 130°C/50 hours ageing. At 160°C/40 hours of ageing, peak yield strength (YS) and tensile strength were recorded and then declined with protracted ageing. Finer grains were also obtained at 160°C/40 hours of ageing, indicating that the appropriate artificial single ageing treatment of the alloy is at 160°C for 40 hours.

Shi and Mahadevan (2001) affirmed that the fatigue failure process stimulated by pitting corrosion of aluminium is comprised of seven stages. They pit nucleation, growth of the pit, progression from pit growth to tiny crack, tiny crack growth, progression from tiny crack to an elongated crack, elongated crack and fracture. The deterioration progression due to pitting corrosion fatigue was examined and the corrosion fatigue lifetime was estimated through probabilistic analyses in the perspective of the damage tolerance approach by the authors. Pitting developing into fatigue crack nucleation and crack growth was regarded to be among the most important damaging mechanisms in ageing structures, although it was concluded that some mechanisms are still not well understood.

According to Jariyaboon et al. (2012), "The effect of pressurized steam can be highly ferocious on the microstructure and oxidation behaviour of aluminium and its alloy". A commercial pressure cooker produced from industrially unadulterated aluminium (AA1050) containing 99.5% aluminium composition was exposed to pressurized steam at a peak temperature of 116°C for ten minutes. The morphology of the surface was examined using FIB-SEM and SEM-EDX, while the compositional depth profiling and phase identification were carried out with the use of GDOES and XRD, respectively. Also, oxidation behaviour was studied via potentiodynamic polarization. The authors observed the generation of a 590 nm thin oxide layer of boehmite on AA1050, associated with particle fall-off and partly dissolved iron. Although an increase in oxidation resistance of the aluminium occurred on exposure of the cooker to pressurized steam, the pressure effect could aggravate the deterioration of the metal on constant exposure.

Sriraman and Pidaparti (2010) affirmed that pitting-triggered damage is majorly responsible for the deterioration of numerous metallic materials, and it primarily affects structural integrity. Also, the process of pitting and crack initiation was considered to govern the entire life of components and structures, especially at nominal cyclic stress. A simple model that considers the consequence of cyclic stress of aluminium alloy 2024-T3 under pitting conditions was developed. The developed model revealed that crack initiation could occur speedily at high stresses even from comparatively small pits. The comparison of the crack initiation life prediction with the existing experimental data suggested that the pit formation, pit growth, crack formation, and cyclic stress are likely factors that determine the stress level on the material.

Sielski (2008) stated that the technology required for the design, fabrication, and maintenance of aluminium structures needs improvement. According to the author, "distortion and residual stresses associated with joining structures made of aluminium have merit and demerit compared to steel". The modulus of elasticity of aluminium is three times less than that of steel. However, the coefficient of thermal expansion is two times higher, which indicates that the strains that take place from the cooling of joints or welds and the environment will generate lower stress in aluminium. The low modulus of elasticity of aluminium suggests that when residual stresses take place, they tend to distort much more than steel. The vibration was also considered to be more acute in aluminium than steel due to the greater possibility of aluminium undergoing fatigue damage. More so, aluminium was regarded as a metal that is prone to fatigue crack propagation and corrosion, and once initiated, it tends to be rapid and concentrated, which requires an immediate approach to restore structural integrity. In conclusion, the author prescribed proper design and maintenance as remedies for minimizing the challenges faced by aluminium structures.

According to Li et al. (2018), "Surface quality is one of the most significant factors that influence pitting corrosion of aluminium structures". The authors emphasized the harmful effect and unpredictability of pitting corrosion and ranked it "the most damaging of all corrosions", as a result of its ability to cause sudden fracture or catastrophic failure. An accelerated corrosion test was carried out on aluminium 2A12 in the laboratory to examine the effect of surface quality on the incidence and progression of pitting corrosion. The outcome of the experiment at the initial stage of corrosion shows that a rough surface is more prone to pitting corrosion and possesses a large corrosion velocity, which increases with time. However, the effect was discovered to disappear gradually on the extension of time, and surface quality was conclusively considered as a key influence on the pitting corrosion of aluminium.

Do Lee (2018) affirmed that the presence of micro-voids is important in determining a material's mechanical properties. With the intention to estimate the dependence of the tensile properties on micro-porosity of A356 aluminium alloy, the tensile properties were determined, and the micro-porosity was examined via SEM. The material was also subjected to solution treatment and artificial ageing. The author discovered that the material subjected to artificial ageing exhibited proportionate tensile properties and micro-porosity, i.e., the ultimate tensile strength (UTS) and micro-porosity decrease with increasing ageing time. The decrease in the UTS was attributed to the micro-void created as a result of the ageing conditions. However, the solution treatment was found to have little effect on the tensile properties and micro-porosity of the material.

9.4 ALUMINIUM CORROSION CONTROL STRATEGIES AND PROCESSING TECHNIQUES

Corrosion control is accomplished via the understanding and discovery of corrosion mechanism, metal type, designs, and corrosion-mitigating materials, and by using protective devices, systems, and treatments. The corrosion deterioration effect on material performance is often reduced with effective control of the environment and error-free designs. The environment is one of the main factors that influence corrosion since variables such as temperature, humidity, and time of exposure initiate corrosion in most environments (Soares et al., 2009; Chauhan & Sharma, 2019). Some reactions involving gases and metal in the immediate surroundings also instigate corrosion. These unwanted reactions could be curtailed by taking appropriate measures to control the environment. This can be reduced by minimizing exposure to saltwater or rain, or by adopting extra multifaceted approaches, such as controlling the quantity of chlorine, sulphur, or oxygen in the surroundings. Treatment of water in boilers with softeners to alter alkalinity, hardness, or oxygen content before coming in contact with metal and material design modification are good protective measures (Kim et al., 2017; Skoczko & Szatyłowicz, 2019).

Design modifications can aid corrosion reduction and improvement of the durability or stability of every active protective anti-corrosive device or system. Preferably, designs should be kept away from dust and water traps. Crevices should also be avoided for materials' durability (Fang et al., 2020a). Corrosion prevention and control does not only entail the characteristics of a structured design to prevent or minimize corrosion but also include the selection of materials, non-destructive examinations for detection of corrosion, cleaning and washing materials, repairs, finishes, and other maintenance activities (Staley, 2017; Fang et al., 2020b).

Fundamentally, the impurities in some metals contribute largely to their poor resistance to corrosion. The presence of anodic and cathodic regions on metals and their ability to react with water and oxygen speeds up the rate of transformation of the metal atom to ion by the loss of electrons (Zhao et al., 2020). Today aluminium is one of the low-cost and valuable metals in the world. The great demand for aluminium is due to its suitability for different applications, which makes it unique compared to other metals. Besides the use of aluminium for the manufacture of cookware, they are also used in diverse fields of engineering constructions, aerospace, and automobiles (Koli et al., 2015; Fayomi & Akande, 2019). Mechanically, chemically and physically, aluminium is a metal similar to brass, zinc, steel, copper, titanium and lead. All these metals can be melted, cast, and machined in a similar way. Apart from a few limitations of aluminium, it is especially desired for the fabrication of structural materials and kitchen appliances due to some exceptional properties, which are as follows:

- Natural generation of protective thin films of coating which isolate the metal from the surrounding.
- Lightweight, about one-third lighter than steel of the same sample dimension.
- Excellent heat and electrical conductivity, almost twice as conductive as copper.
- Reasonable ductility, low density, and melting point.
- Non-magnetic, making it valuable for electrical shielding as in computer disks, dish antennas, and bus bars.
- Outstanding shock and sound absorber make it reliable for auto bumpers ceilings construction, respectively.
- Recyclability, 100% recyclable and indistinguishable from the original product.
- Cost-effective for production runs, re-melting needs modest energy.
- Non-sparking does not produce sparks in contact with itself or other materials.
- Increase in tensile strength as temperature reduces, unlike steel that hastily goes brittle at low temperatures.
- Reasonable tensile strength and hardness, but exhibit lower strength compared to stainless steel (Tisza & Czinege, 2018; Fayomi et al., 2019).

Aluminium can be hardened with the inclusion of alloying elements like magnesium, copper, and other elements followed by heat treatments; precipitation hardening; or age hardening, which are the most frequently used techniques for strengthening metal alloys. Pure aluminium and aluminium alloyed mostly with magnesium or manganese is not affected by heat treatment. However, aluminium alloys containing zinc, copper, or a blend of silicon and magnesium respond favourably to heat treatment (Stemper et al., 2019). High-purity or non-alloyed metals can majorly be hardened by grain size reduction and strain hardening. While heat treatment does not enhance the properties of all aluminium alloys, heat treatment of several alloys increases the ease of forming or the strength of the end product. Unlike steel, aluminium needs rigid heat control to accomplish the most favourable results. Thus, special equipment is normally required. All alloys are strengthened by cold working techniques such as wire drawing or cold rolling. The strength of aluminium alloys can be improved via different combinations of cold working, heat treatment, and alloying (Gu et al., 2016).

The use of aluminium in raw form to produce cookware is certainly undesirable due to its extreme softness and reactive nature (Vu et al., 2019; Sim et al., 2020). Hence, it must be processed and anodized. Anodized aluminium is the main aluminium cookware in existence, and these wares are produced through treatment with chemicals to harden and reduce its reactivity. Anodizing aluminium simply involves an increase in the thickness of the natural oxide layer of the surface through an electrolytic passivation process or application of anodic films. By anodizing, the part to be treated acts as the anode of the electrolytic cell; hence, enhancement in wear, corrosion performance, and better adhesion for paint primers are achieved (Elkilany et al., 2019). Anodizing influences the near-surface crystal structure and the surface microscopic texture. However, anodized aluminium exhibits a harder surface than bare aluminium but has wear resistance ranging from low to moderate. Improvement in the wear resistance can be attained by increasing the thickness of the anodic films, which is generally much stronger than metal plating and most paint types but also more brittle. Thus, for this reason, anodic films are less liable to crack and peel, which could result from ageing and wear, but more prone to cracking that is due to thermal stress (Abdel-Gawad et al., 2019; Roshani et al., 2017).

9.5 HIGH ENTROPY ALLOYS (HEAS): CHARACTERISTICS, PROPERTIES, AND APPLICATIONS

Humans advanced their ability to formulate materials such as alloys that evolved from simple to complex compositions, consequently improving functions and performances and promoting the advancements of human civilization. Numerous types of alloying elements have been used as reinforcement to enhance the thermal, mechanical, and electrochemical properties of engineering materials. However, there has been a paradigm shift to HEAs in recent times, due to their inherent microstructural properties and many other impressive properties, such as high hardness and strength, outstanding corrosion resistance, fatigue, fracture, thermal stability, and irradiation resistance, in terms of which they engulf the conventional alloys with several potential applications (Zhang et al., 2018). HEAs have been used for barely above a decade and have stimulated new ideas and also encouraged the investigation of the enormous composition space presented by multi-principal element alloys (Miracle & Senkov, 2017). Researchers have shown that HEAs have immeasurable potential in material development. They are types of metallic alloy system that comprises near equiatomic concentrations of four or more alloying elements ranging from 5 to 35 at.% (Park et al., 2016; Takeuchi et al., 2018; Gwalani et al., 2019).

The development of HEAs with properties that will replace the conventional aluminium alloy used for the production of cookware is envisaged due to reduced defects, low cost, and high efficiency. It is produced through various manufacturing techniques, which has made it attractive, increasing attention in recent years (George et al., 2019; Lyu et al., 2019; Dobbelstein et al., 2018). Some of the HEAs manufacturing techniques are arc melting, Bridgman solidification, mechanical alloying, sputtering, thermal spray, induction melting, laser cladding, electrodeposition, spark plasma sintering (SPS), and PVD (Ji et al., 2015; Cieslak et al., 2018; Jiang et al.,

2019). HEAs as a novel class of materials have several existing potential applications, including submarines, state-of-the-art race cars, spacecraft, jet aircraft, nuclear reactors, long-range hypersonic missiles, and nuclear weapons (Prabhu et al., 2019). Furthermore, studies have established that several HEAs have significantly improved strength-to-weight ratios, with a superior measure of tensile strength, fracture resistance, as well as oxidation and corrosion resistance than the usual alloys (Youssef et al., 2015; Yim et al., 2019; Ganesh & Raghavendra, 2020). The mechanical properties of any HEAs are a function of its crystal structure. Body-centred cubic (BCC) HEAs characteristically have low ductility and high YS, whereas FCC HEAs possess high ductility and low YS.

The HEAs were proposed originally for the benefit of phase stabilization through entropy maximization (Li et al., 2016). Some of the currently existing HEAs such as CoCrFeMnNi, AlxCoCrFeNiTi ($0 \leq x \leq 2$), VNbMoTaW, AlCrCuWZr, AlBFeNiSiNb, AlLiMgScTi, AlCoCrFeNi, Al0.5CoCrCuFeNi, Al*x*Co15Cr15Ni70-*x* ($0 \leq x \leq 35\%$), and HfNbTaTiZr have particularly been well-known for their outstanding mechanical and microstructural characteristics (Miracle & Senkov, 2017). VNbMoTaW was used as a refractory alloy, and it was able to maintain a high YS greater than 600 MPa even at 1400°C and this outperformed the regular superalloys such as Inconel 625 with an YS of 430 MPa at 1065°C and also reasonably showed superiority in mechanical properties compared to Inconel 718. Up to date, a lot of HEAs with promising mechanical and chemical have been reported, e.g., Al0.2Co1.5CrFeNi1.5Ti and Co1.5CrFeNi1.5Ti alloys with high wear resistance; high-strength NbMoTaV at high temperatures; and BCC AlCoCrFeNi at ambient temperature. Also, Cu0.5NiAlCoCrFeSi generally has corrosion resistance that is much better than commonly used 304-stainless steel (Zhang et al., 2014; Akca & Gursel, 2015; Shang et al., 2018; Hache et al., 2020).

HEAs like CoCrFeMnNi exhibited excellent mechanical properties and high fracture resistance at low temperatures with increasing YS and ductility at a reduced temperature, from room to 350°C. This was accredited to the origination of nanoscale twin boundary, a supplementary mechanism of deformation that was not present at elevated temperatures. HEAs may be found useful for low-temperature structural material applications due to high toughness or energy-absorbing material applications (Otto et al., 2013). However, successive research has shown that alloys of lower entropy with smaller numbers of elements or non-equiatomic compositions could have higher toughness or strength (Wu et al., 2014; Zaddach et al., 2015). Higher endurance and fatigue limit compared to conventional titanium alloys and steel was observed with Al0.5CoCrCuFeNi HEA, while samples of NbTaMoW exhibit astonishingly high YSs between 4 and 10 GPa, worthwhile ductility and high thermal stability, thus termed a material for potentially high temperature and stress application (Zou et al., 2014).

Most engineering material often deteriorates in properties on exposure to high temperatures as revealed by several researchers (Choi et al., 2019). The effect of high-temperature exposure ranging from 650°C to 900°C on Al5Ti5Co35Ni35Fe20 HEA's mechanical and microstructure properties studied by Bala et al. (2020) unveiled that at 650°C temperature exposure, the utmost UTS and YS of 1370 and 1050 MPa, respectively. Rising the temperature further lowered the mechanical properties.

More so, the air quenching of Al5Ti5Co35Ni35Fe20 HEA initiates gamma (γ) precipitation, which was dispersed homogeneously all through the microstructure. At high-temperature exposure, the growth of gamma (γ) particles was noticed, and at temperatures above 700°C, more precipitation of gamma (γ) particles was observed. Generally, the properties of HEAs can be modified by replacing some of the compositional elements. More often than not, at elevated temperatures, refractory HEAs were found to possess remarkable strength but were brittle or frail at room temperature. However, HfNbTaTiZr HEA demonstrated the reverse, but its strength at elevated temperatures was inadequate. In an attempt to improve the strength of the HEA at high temperatures, it was modified to HfMoNbTaTiZr and HfMoTaTiZr alloys with a simple BCC structure. The outcome of the study showed that HfMoNbTaTiZr had YS that is six times greater than HfNbTaTiZr at 1200°C with the retention of 12% fracture strain at room temperature (Juan et al., 2015).

Overwhelmingly, the high mixing entropy effect of HEAs promotes the formation of chaotic or disordered solid solutions of the hexagonal closed pack (HCP), FCC or BCC arrangement, as a replacement for complex intermetallic compounds (Gao et al., 2016). The arbitrary structure of numerous elements in the solid solutions of HEAs is often able to create a locally disordered chemical environment, which is usually the reason for unique corrosion-resistant properties (Diao et al., 2015). The corrosion behaviour of HEAs examined in a variety of aqueous environments such as acid, saline water, and high-pressure and high-temperature water showed that the synergetic or combined effect of passivating elements such as Ta, Ti, Zr, Cr, Mo, Ni, etc. in HEA systems provided better corrosion-resistant characteristics compared with the traditional alloys (Qiu et al., 2015; Liu et al., 2015; Shi et al., 2019). Furthermore, the outstanding characteristics such as improved strength-ductility, enhanced fatigue resistance, high thermal stability, and high fracture toughness make HEAs attractive in several manufacturing fields, particularly as advanced structural alloys, which are enormously demanded in intense service environments such as in the aerospace, turbine, and nuclear industries (Santodonato et al., 2015; Seifi et al., 2015; Li et al., 2016).

The HEAs have yielded significant advancements in the field of technology. The core effects and unique properties of HEAs are responsible for the random solid solution (RSS) found in them. As HEAs become increasingly acceptable, it has evolved into a well-known crystalline system of four or more elements, each comprising 5–35 at.% with a structure of at least one unique RSS phase. The wide application of HEAs reduces single-phase, multi-phase, and complex-engineered phase as the three main classifications. These classifications permit the precise analysis of HEA's exceptional properties as the field persistently grows. The development of experimental laws, thermodynamic factors, and modelling through computation are proving to be essential to the detection of more promising HEAs. The outstanding momentum of HEAs field is expected to persist, with the research progressing in the direction of the relevance of these materials for the diversity of engineering designs (Macdonald et al., 2017). Innovative ideas and the new route of alloy research can be continually expanded by investigating the structure, organization, and properties of HEAs since they were considered a major advancement in alloying theory in recent decades (Divinski et al., 2018; Poulia et al., 2019).

9.6 CONCEPT OF HEAs

HEAs are generally known to contain at least four key elements (Lu et al., 2020; Wang et al., 2021). Chemical composition increases with an increase in principal elements. This implies that HEAs exhibit a larger chemical composition compared to others. Understanding the possible interactions that exist between the phases and mixing entropy is a vital subject of concern for manufacturers in the production of HEAs. Due to this concern, researchers have asked a lot of questions about the mixing entropy in HEAs. Consequently, the entropy of mixing per mole could be written configurationally, as shown in Equation (9.1).

$$\Delta \text{Smix} = -Kg \sum_{i=1}^{\acute{\text{E}}} \acute{\text{M}}\text{i} \ln \acute{\text{M}}\text{i} \tag{9.1}$$

where Kg is the gas constant, Ḿi is the ith element's molar fraction and É is the number of elements or metals contained in the mix. Contour diagrams of ΔS_{mix} for compound alloying systems have shown that the central region is typically the high entropy section, where ΔS_{mix} is much greater, while for the traditional alloys ΔS_{mix} is lesser, and is signified at the corner sections. In the majority of metal alloys, the energy gained is usually enough to stabilize the entropy of a random solid solution phase against intermetallic compounds. Therefore, for equiatomic HEAs, ΔS_{mix} is written in the form shown in Equation (9.2).

$$\Delta \text{Smix} = Kg \ \ln \acute{\text{E}} \tag{9.2}$$

Due to the high entropy effect, the stability of the arbitrary solid solution against the intermetallic compounds may well be realized when numerous elements are combined in an equimolar fraction (Guo, 2015; Ye et al., 2016; Park et al., 2016, Fu et al., 2017). Researchers have also shown that HEAs usually seem to contain tremendously lesser phases than Gibb's phase rule would allow. There is an assumption that configurational entropy offers a major stabilizing effect even in alloys of several phases. Also, a lot of studies could not substantiate phases available in HEAs; thus, it has become essential to cautiously assess the stability before the number of phases can be recognized. Though substantiating the absolute thermodynamic stability of alloys has been regarded experimentally impossible, nevertheless, in the estimation of the stability of HEAs, there are significant characteristics that the studies ought to have. Appropriate heat treatments ought to be selected for the microstructural homogeneity of the as-cast alloy and encouragement of phase decomposition. Near-stable microstructures are produced by heat treatment of as-cast HEAs and the required confirmation of their stabilities can be made through careful characterization. Most research has shown that most HEAs are made of BCC, FCC, or a combination of the solid solutions of both due to their high configurational mixing entropy. The configuration of these HEAs is organized to accomplish potential attributes such as high corrosion, oxidation, and wear resistance as well as hardness and ductility (Pickering & Jones, 2016; Nong et al., 2018; Joseph Park et al., 2019).

9.6.1 Four Core Effects of HEAs

The characteristics of HEAs have been examined by several authors. There are indications that the fundamental concerns such as the thermodynamic root have not been completely resolved. Nevertheless, HEAs are reported to demonstrate superior mechanical properties, outstanding physical characteristics, super-high fracture toughness, and corrosion resistance superior to that of many alloys. Their outstanding strengths are only comparable to some structural ceramics. They also possess exceptional resistance to corrosion and high conductivity (Li & Zhang, 2016; Lyu et al., 2018). The astonishing properties of HEAs have been largely attributed to the following:

i. The effect of kinetic sluggish diffusion or slow dispersion, which slows down the growth of the nucleus of the second phase from a single-phase solid solution, thus aiding the formation of nano-precipitates.
ii. The stern deformation or distortion effect of the structure lattice, which provides extra strength and thus assists the slow kinetics in HEAs.
iii. The properties of cocktail or mixture effect, which could improve the characteristics through alloying.
iv. The effects of high entropy, which offers the thermodynamics for the stability of the single-phase solid solutions.

These descriptions not only explain the complex happenings that await clarification but also represent important procedures for examining the configuration and behavioural association in HEAs (Zhang et al., 2014; Li et al., 2018).

9.7 VARIOUS TECHNIQUES OF PRODUCING HEAs

HEAs are found complex to produce using the existing methods a few years back, and characteristically call for both costly materials and speciality of processing procedures. HEAs are predominantly manufactured using techniques that depend on the phase of metals—provided the metals are combined while in a gaseous, solid, or liquid state. In recent times, most HEAs have been manufactured through liquid-phase techniques such as Bridgman solidification, induction melting and arc melting. Generally, solid-state processing is carried out by mechanical alloying employing a ball mill of high energy. This technique generates powders that can further be processed by the traditional powder metallurgy techniques or SPS. This technique allows for the production of alloys not easy or unfeasible to manufacture using casting. Gas-phase processing involves molecular beam epitaxy or sputtering that can be utilized carefully to control various elemental compositions to obtain high-entropy metallic films. HEAs have also been produced using other techniques such as electrodeposition, laser cladding, and thermal spray. In the manufacturing of HEAs from mechanical alloying, the milling of essential constituents is involved. Two phases known as terminal phases are often contained in the product. These phases are the amorphous and solid solution phases. To avoid welding or bonding of powder, the process of milling can first be done in the presence of liquefied nitrogen. Several

HEAs are produced from cold casting, but instead of cast and melting, mechanical alloying is utilized for alloys preparation where numerous elements in the mixture exhibit high vapour pressure (Yu et al., 2016; Senkov & Miracle,2016; Erdogan et al., 2018; Wang et al., 2018; Laplanche et al., 2018; Wei et al., 2018; Tian et al., 2019; Yin et al., 2019; Sarswat et al., 2019; Vaidya et al., 2019; Kim et al., 2020).

9.7.1 Arc Melting

Arc melting is carried out using an electric arc furnace, which heats the charged materials using an electric arc. The size of an industrial arc furnace ranges from about one ton unit to approximately 400 tons depending on the area of application. Most industrial electric arc furnace functions at a maximum temperature of about 2073 K (1800°C), whereas laboratory unit furnaces can function above the temperature of 3273 K (3000°C). The major difference between the arc furnaces and induction furnaces is that the materials charged into the arc furnaces get exposed directly to the electric arc and the furnace terminal current goes through the materials charge. Electric arc melting is commonly used, and in the process, arc melting is carried out in a closed argon atmosphere and then the casting of the molten alloy follows. One major drawback of heating elements such as silver and copper simultaneously is their ability to form a hypoeutectic, which makes an effort to separate from other constituents. For this reason, the sintering of alloys or elements also known as powder metallurgy can be used. The use of an electric arc furnace allows alloys to be produced from scrap metal feedstock. This significantly minimizes the required energy to produce alloys when compared with the principal ones produced from ores (Cieslak et al., 2018; Zhang et al., 2019; Hou et al., 2019).

9.7.2 Powder Metallurgy

Powder metallurgy has been used for the production of HEAs and components from metal powders. The processes can significantly minimize or prevent the necessity for the use of metal removal processes. This considerably reduces losses of manufacturing yield and production costs. Powder metallurgy is also often employed to produce unique components that are not possible to produce from other forming or melting techniques. It is widely used for the industrial manufacturing of varieties of tools. The powder metallurgy press and sinter procedure usually consists of pulverization (powder blending), compaction of die, and sintering. Die compaction is commonly carried out at ambient temperature. The high-temperature process of sintering is generally performed at atmospheric pressure. More so, the atmospheric composition is carefully controlled. Secondary processing such as heat treatment regularly follows to achieve improved precision or particular characteristics (Cao et al., 2018; Zhang et al., 2018; Sekhar et al., 2019; Torralba et al., 2019; Guo et al., 2019).

9.7.3 Stir Casting Technique

In a stir casting process, mechanical stirring aids the distribution of reinforcing phases into a molten matrix. In the late 1960s, stir casting of MCCs was initiated by

S. Ray. Alumina particles were introduced into aluminium melted by stirring molten aluminium alloys containing fine particles of ceramic. Stirring the furnace mechanically is the most important feature of this process. The resulting molten alloy, with the particles of ceramic, can then be employed for sand casting, die casting, or permanent mould casting. Stir casting is appropriate for the development of composites with up to 30% weight fractions of reinforcement. To minimize porosity, refine microstructure and homogenize the reinforcement distribution of the cast or developed composites, the casts are sometimes further extruded. A key challenge associated with the stir casting route is the segregation of reinforcing particles, caused by the settling of these particles during the processes of melting and casting. The ultimate particle distribution of the solid depends on material properties and process parameters such as the melting temperature, rate of solidification, relative density, mixing strength, and wetting state of the particles. The particle distribution in the molten matrix depends on the stirring parameters, shape and size of the mechanical stirrer, properties of the added particles, and position of the mechanical stirrer in the melt. The two-step mixing process is a recent development in stir casting, where the matrix material is liquefied by the application far above its melting temperature. The liquid metal is then allowed to cool to a temperature between the solid and liquid points and kept in a semi-solid state. In this phase, the preheated particles are added and stirred. The slurry is once more heated completely to a liquid state and stirred vigorously. This two-step mixing technique has been used in the manufacture of aluminium. Stir casting is the most economical and entrenched MCC fabrication technique, and for this reason, it is presently the most admired commercial means of manufacturing aluminium-based composites.

9.7.4 Induction Melting

Induction melting is carried out using an induction furnace. It is an electrical furnace that allows the application of heat through the induction heating of metals. An induction furnace has been used to melt aluminium, copper, iron and steel, and several metals. The induction furnace has the advantage of a clean, well-controllable process of melting and is energy-efficient compared to most other means of metal melting. An induction furnace is used in most contemporary foundries. In recent times, cupolas used in most iron foundries have been replaced with induction furnaces to melt metals, as the former furnace releases plenty of dust and several pollutants. Since combustion or arc is not used, the material's temperature is not higher than the melting temperature, thus the loss of important alloying elements can be prevented. Another advantage of induction melting is that the heat is produced inside the furnace, and it charges itself being charged by applying combustion fuel or any other external source of heat. This can be vital in applications with the issue of contamination. The main challenge to the usage of induction furnaces in foundries is the lack of refining ability and as such, loss of some alloying elements could occur due to oxidation. Therefore, before charging materials to the furnace they must be clean/free of oxidation products (Laurent-Brocq et al., 2015; Patil & Ghatge, 2017; Karlsson et al., 2019; Wang et al., 2020).

9.7.5 Bridgman Solidification

Bridgman solidification is a technique of developing single-crystal ingots using a Bridgman furnace. The technique is executed by heating polycrystalline material in a crucible above its melting point and gradually cooling it from the seed crystal location end. Single crystal material is produced progressively along the crucible's length, which can either be of vertical or horizontal geometry. Within a Bridgman solidification furnace, there exists a distinctive temperature gradient, and a crucible enclosing the material is transported slowly along the axis of the temperature gradient. As the material liquefies through declining temperatures it produces a single crystal. In a similar method known as Stockbarger, a name frequently used interchangeably with the Bridgman technique, the furnace contains three different temperature regions. The topmost region has a temperature above the melting point while the bottom or base region has a temperature below the melting point. There is a middle region, which functions as a baffle between the upper and the bottom region. This technique is often employed in a Bridgman furnace because it provides better control over the temperature gradient. The Bridgman technique is an admired option for the production of some semiconductor crystals, which can be complex to produce using other production techniques (Yang et al., 2015; Hallensleben et al., 2017; Wang et al., 2018; Feuerbacher et al., 2018).

9.7.6 Mechanical Alloying

Mechanical alloying is a powder processing and solid-state technique for producing alloys. It involves continuous cold welding, rupturing, and re-welding of mixed powdered particles in a ball mill of high energy to generate a homogeneous material. Mechanical alloying has been shown to have the ability to produce alloy phases of different equilibrium and non-equilibrium beginning from pre-alloyed powders or blended elements. The non-equilibrium phases such as quasi-crystalline and metastable-crystalline phases, supersaturated solid solutions, amorphous alloys, and nanostructures have been synthesized. However, care must be taken to avoid powder or particle contamination. Mechanical alloying is similar to metal powder processing, where the blending of metals may be carried out to fabricate superalloys. Three steps are involved in mechanical alloying. In the first step, the alloying elements are blended in a ball mill and crushed to a fine powder. This is followed by the application of a hot isostatic pressing process to compress and sinter the powder simultaneously. Heat treatment is the final stage, which aids in the removal of internal stresses generated during any cold compression. Mechanical alloying is often suitable for producing alloys for aerospace components and turbine blades (Murali et al., 2016; Pan et al., 2018; Xie et al., 2018; Cheng et al., 2019).

9.7.7 Spark Plasma Sintering

The SPS has a major characteristic of the un-pulsed and pulsed AC or DC, which directly passes through the die and the powder compact. This is common in the case of conductive material samples. Heating plays a principal role in powder compact densification and often leads to the achievement of near hypothetical density at a reduced sintering temperature compared to traditional sintering methods. In SPS, the heat is generated internally. This is in contrast to the traditional hot pressing, where the heat is supplied by outer heating constituents. The internal heat generation facilitates a high heating rate, making the sintering process extremely fast. The speed of the process enables the quick densification of powders with nanostructure or nano-size, hence preventing coarsening of materials. This has made the SPS a good technique for producing alloys with improved thermoelectric, magnetic, piezoelectric, and optical properties (Kumar et al., 2018; Waseem et al., 2018; Kumar et al., 2018; Vaidya et al., 2019).

9.8 A REVIEW ON PRODUCTION OF SELECTED HEAs AND THEIR PROPERTIES

Senkov et al. (2011) studied the mechanical properties of V20Nb20Mo20Ta20W20 and Nb25Mo25Ta25W25 refractory HEAs. These alloys were prepared through vacuum arc melting by ensuring the corresponding elements were mixed in equal molarities. High-purity titanium was used as a getter in a high chamber, for residual gas. The buttons were re-melted four times, and each was stirred for about one hour in a liquid state, to achieve a homogeneous mixture and distribution of elements. The buttons exhibited a shining surface, which indicates that oxidation did not take place during the vacuum arc melting. Neutron diffraction analyses of the alloys at room temperature and annealed at 1400°C showed that both alloys had a single-phase and BCC crystal structure. In the process of deformation at ambient temperature, V20Nb20Mo20Ta20W20 alloys demonstrated considerably high yield stress of 1246 MPa, but limited ductility of 1.5% of plastic strain. Likewise, Nb25Mo25Ta25W25 showed a relatively high yield stress of 1058 MPa and low ductility of 1.5% of plastic strain. Above the 600°C–1000°C temperature range, the two alloys showed good plastic flow with compressive strain going above 10%–15% before localized shear led to a drop in strength, which indicated that the brittle-to-ductile transition temperature for the two alloys is situated between ambient temperature and 600°C. Nb25Mo25Ta25W25 alloy possesses yield stress of 561 MPa at 600°C and diminishes down to 405 MPa at 1600°C as the temperature increases. V20Nb20Mo20Ta20W20 alloy was approximately 200 MPa stronger in the 600°C–1200°C temperature range. However, its yield stress declined more swiftly than that of the Nb25Mo25Ta25W25 alloy. Microstructural examination of the two alloys indicated that grain boundary sliding took place at these temperatures, which was responsible for the formation of void and fracture along grain boundaries. In summary, these alloys showed a drop in yield stress by 30%–40% between ambient temperature and 600°C but were comparatively not sensitive to a temperature above 600°C, which compares favourably with conventional superalloys.

Chuang et al. (2011) investigated the wear and microstructural characteristics of AlxCo1.5CrFeNi1.5Tiy HEAs; ($0\leq x\leq 0.2$) and ($0.5\leq y\leq 1$). The alloy ingots were prepared by arc-melting through the combination of the component elements, which had above 99.5 wt.%. This was carried out in an argon ambience. The as-cast alloys were air-cooled after being homogenized for four hours at 1100°C. The alloys were further aged for ten hours at 800°C and air-cooled. The microstructural assessment of the specimens was conducted using standard metallographic equipment (XRD and SEM). The wear property of the alloys was examined by pin-on-disk sliding per ASTM G99 standard in a dry sliding situation. Due to the good mechanical characteristics of AISI M2 high-speed steel, it was used as a reference material in the wear test. The wear test samples were of cylindrical shape of diameter 8 mm and height 3 mm. Each of the test samples was held rigidly in a holder to wear against an AISI M2 steel disk of diameter 75 mm and hardness value of 890 HV. The gap between the samples and the centre of the disk was 20 mm, sliding speed was 0.5 m/s with a load of 29.4 N. The results of the experiment revealed that the quantity of Ti and Al robustly influences the microstructure and predominantly the morphology of the phases. Co1.5CrFeNi1.5Ti and Al0.2Co1.5CrFeNi1.5Ti alloys were also found to have wear resistance that is more than two times superior to conventional wear-resistant steels with comparable hardness. The exceptional wear resistance of these HEAs was ascribed to their excellent resistance to thermal softening and outstanding anti-oxidation properties. However, the melting process was repeated many times, beyond expectation to enhance the microstructural homogeneity of the ingots.

Hemphill et al. (2012) examined the fatigue behaviour of Al0.5CoCrCuFeNi HEA produced by the arc melting technique. Samples of the alloy were prepared by arc melting the compositional elements in a water-cooled copper hearth at a current of 500 amperes. All the constituent elements have a purity of at least 99 wt.%, and melting was achieved in a vacuum of at least 0.01 T. To improve the element microstructural homogeneity of the samples, their liquefaction and solidification were repeated more than five times. The cast samples were subjected to annealing at a temperature of 1000°C for six hours, quenched with water and cold-rolled to 84% reduction, with a thickness of 3 mm. The rolled sheets were then machined to fatigue samples of a dimension 25 × 3 × 3 mm. The microstructural behaviour of tensile sections of the fatigue samples was examined, using backscattered electron microscopy. The fractured surfaces were studied using SEM, to determine the fatigue mechanisms. The fatigue endurance limit (FEL) of Al0.5CoCrCuFeNi alloy was ascertained to be between 540 and 945 MPa, and the ratio of FEL to UTS was between 0.402 and 0.703. The result from this study was adjudged better than many conventional alloys. Although the corrosion and wear characteristics of the material were not examined.

Wang et al. (2014) investigated the mechanical, phases, and microstructural properties of AlxCoCrFeNi HEAs; ($0 \leq x \leq 1.8$) at elevated temperatures. The ingots of HEA with noticeable composition AlxCoCrFeNi ($0 \leq x \leq 1.8$) in molar proportion were prepared using the vacuum arc re-melting technique. All the raw elements had at least 99.9% purity. Materials of approximately 200 g were melted under a particularly high-purity argon environment in a water-cooled mould made of copper to avoid oxidation. Re-melting of each ingot was done at least five times to boost chemical microstructural homogeneity. The solidified ingots were of dimension 45 × 45 × 10 mm. Crystal structures of the cast ingots at high temperatures were studied, utilizing a high-temperature X-ray diffractometer (HT-XRD) with CuKα1+

2 radiations functioning at 30 kV/20 mA. The phase transition and melting point of the alloys were analysed with the aid of a high-temperature differential scanning calorimeter at a heating speed fixed at 10 Kelvin/min from ambient temperature to 1723 Kelvin. The hot hardness of the alloys was measured using a high-temperature Vickers hardness tester (Model AVK-HF). For a good understanding of microstructure at elevated temperature, five representatives of Alx alloys (Al0.3, Al0.5, Al0.7, Al0.9, and Al1.5) were heat-treated at 1173 and 1373 K in the atmosphere for two hours and quenched in water. The quenched samples were polished, and the faces were then observed under a field-emission gun SEM. The chemical composition of samples was investigated with SEM equipped with EDS, operating at 15 kV and thin foils were used for careful phase recognition under transmission electron microscopy, operating at 200 keV. After completion of the experiment, structural evolution with temperature was classified into four, namely: FCC structure for alloys with (Al0–Al0.3), mixed structure (FCC + ordered BCC) for (Al0.5–Al0.7), ordered BCC + disordered BCC structure at (<873 K) and mixed structure (FCC + ordered BCC) at (≥873 K) for (Al0.9–Al1.2) and ordered BCC structure for (Al1.5–Al1.8). The hot hardness transition temperature (Tt) range of Alx alloys was at 810–930 K and this is about 0.5Tm. Al0 and Al0.3 alloys exhibit the peak softening coefficient below Tt and the least softening coefficient above Tt. Al0.9 and Al1.0 alloys possess the utmost softening coefficient above Tt but re-hardened between 973 and 1073 K. Softening difference was ascribed to variation in constituent phases, distribution of phase, and morphology.

Xiao et al. (2014) studied the mechanical, microstructural, and corrosion characteristics of AlCoCuFeNi-(Cr, Ti) HEAs. The HEAs were synthesized using non-consumable arc melting to explore the effect of Ti and Cr on their corrosion and mechanical properties. For equiatomic ratio HEAs, including AlCoCuFeNi, AlCoCuFeNiTi, AlCoCuFeNiCr, and AlCoCuFeNiCrTi, were prepared, ensuring that the purity of all constituents was at least 99.9%. The crystalline structures were identified using XRD with CuKα radiations while the microstructure characterization and identification of the chemical composition of different phases were carried out employing SEM and EDS, respectively. Hardness measurements were achieved using a Vickers hardness tester (Micromets of model 5104) under a test load of 50 N and a time of 20 seconds. Compression tests were carried out at room temperature using an Instron 3369 testing machine and at a strain rate of 1×10^{-4} s^{-1}. Fracture strength was measured through the peak stress the material can withstand in the process of compression before breaking. Corrosion properties of the alloys were studied in 0.5 M of H_2SO_4 at 298 and 366 K. The outcome of the investigation revealed that the as-cast AlCoCuFeNi-based HEAs exhibit a dendritic structure (FCC and BCC solid solution phases). Inclusions of Ti and Cr were found to refine the grains of the as-cast alloy, but the segregation of Cu occurred, which led to the formation of the FCC Cu-rich phase. The Ti-free alloy was also discovered to possess a microstructure that is different from Ti-containing alloys. The addition of about 16.7 at.% of Cr and Ti elements in AlCoCuFeNi eases the formation of BCC solid solution and FCC phase, respectively. More so, AlCoCuFeNi alloy exhibited YS, fracture strength, and hardness of 1060 MPa, I452 MPa, and 387 HV, respectively, while the AlCoCuFeNiTi alloy possesses YS and hardness of 1612 MPa and 623 HV, respectively. AlCoCuFeNiCr alloy exhibits fracture strength and hardness of 1857 MPa and 459 HV, respectively. AlCoCuFeNiCrTi possesses YS and hardness of 1612 MPa and 510 HV, respectively. AlCoCuFeNiCr alloy had the highest fracture strength, while AlCoCuFeNiCrTi had the peak YS and hardness, respectively. These results indicated that Cr and Ti inclusion improved the mechanical properties of AlCoCuFeNi alloy. However, the addition of Ti lowers the corrosion resistance of AlCoCuFeNi alloys at 298 and 366 K. The presence of Cr also enhances its corrosion resistance. However, the combination of Cr and Ti in the formulation increased the corrosion rate, and Cr_2O_3 passivation was discovered to break down higher temperatures above 366 K.

Joo et al. (2017) examined the properties and structure of ultrafine-grained (UG) CoCrFeMnNi HEAs produced by SPS and mechanical alloying. A combination of the constituent elements with at least 99.9% purity and particle size less than or equal to 300 μm were used as the starting materials. Mixing of basic metallic powders at the initial stage was done using high-energy ball milling for 20 and 60 minutes inside a planetary ball mill, by means of zirconia balls of 5 mm diameter in argon gas. The ratio of ball to powder was 10:1, and the rotating speed was 1100 rev/min. The milled powders were subsequently consolidated using SPS at 900°C or 1100°C, under 50 MPa uni-axial pressure for eight minutes in argon gas. The SPS temperature, mechanical alloying time, and defect (contamination) were found to have strongly affected the mechanical and final microstructural properties. Nanocrystal FCC was produced during mechanical alloying, which was sustained as the matrix after SPS at 900°C and 1100°C. Nevertheless, there was a transformation of Cr carbides near the surface as a result of carbon contamination. An increase in mechanical alloying time improved the stability of the FCC phase, and there was also an increase in ZrO_2 contaminant from the mechanical alloying balls. The UG microstructure was attained at 60 minutes of mechanical alloying and 900°C of spark plasma sintering. The 900°C SPS samples possess a higher strengthening effect than the 1100°C SPS samples, even though lesser degrees of contamination were required for tensile ductility. This occurrence was attributed possibly to unidentified nanoparticles, which were not detected. Cr and Mn required more time to dissolve into the FCC phases compared to other elements, and the corrosion properties of the alloy were not investigated.

Tsau and Wang (2018) investigated the hardness, corrosion, and microstructural effect of Nb and Mo on FeCoNi alloy by producing FeCoNiNb0.5Mo0.5 and FeCoNiNb HEAs via arc-melting in an argon atmosphere. The microstructural evolution of the alloys was examined using SEM/EDS operated at 15 kV. XRD with CuKα, having a wavelength of 1.541 Å was used for the structural characterization. With the use of a high-resolution TEM operated at 300 kV, lattice images and microstructures of the alloys were obtained. The hardness of the samples was examined using the Vickers hardness method, under a test load of 30 kgf. The corrosion resistance behaviour of the alloys was studied in 1 M H_2SO_4 and 1 M NaCl solutions via a potentiostat/galvanostat (Autolab PGSTAT302N). The polarization results were compared with that of commercial 304 stainless steel. The results of this experiment showed that FeCoNiNb0.5Mo0.5 and FeCoNiNb HEAs exhibit binary-phased dendrite formations. The hardness of FeCoNi, FeCoNiNb, FeCoNiNb0.5Mo0.5 and 304-stainless steel were 112, 798, 629, and 185 HV. These hardness results showed that FeCoNiNb alloy had higher hardness values than FeCoNiNb0.5Mo0.5 alloy, although both alloys possess higher hardness than 304-stainless steel. The resistance of FeCoNiNb0.5Mo0.5 alloy to corrosion was better than FeCoNiNb alloy in 1 M NaCl and 1 M H_2SO_4 solutions, but both alloys have corrosion resistance that is inferior to that of FeCoNi alloy. There was a huge distortion in the lattice due to the existence of elements with larger radii. However, all the alloys offered comparable corrosion resistance to 304-stainless steels in the test solutions.

Tseng et al. (2018) studied the distinctive properties of Al20Be20Fe10Si15Ti35 lightweight HEA produced by vacuum arc welding of the constituent elements with a current of 500 A in a cooled copper mood. Elements with 99.9 wt.% purity were used. The chemical composition of various phases of the alloy was examined using a field emission electron probe microanalysis, while density was obtained using the Archimedes technique. Vickers hardness tester was used to determine the hardness of the alloy under a test load of 5 kg, 50 μm/s loading speed, and a holding time of five seconds. Shimadzu XRD of model 6000 was used to analyse the crystal structure. The XRD was operated at 30 kV and 20 mA with a scan rate of 4°/min from 30° to 90°. The resistance of the alloy to corrosion was tested through the measurement of weight gained by the alloy samples after 1, 5, 10, 50, 100, and 150 hours of annealing at 700°C and 900°C. Three phases (two minor and one major) were noticed in the as-cast microstructure. The alloy density was found to be 3.91 g cm^{-3} and its hardness was 911 HV, which implies that it is harder than quartz. This hardness value and hardness value to density ratio was regarded as the maximum ever reported of any lightweight alloy. The oxidation resistance of the alloy at 700°C and 900°C was considered outstanding because it exceeded that of Ti-6Al-4V. The alloy was conclusively considered a potential material for high-temperature applications, especially where oxidation resistance, lightweight, and wear-resistant components are required. However, it was observed that some of the elements in the intermetallic phases were not soluble enough.

Chauhan et al. (2020) examined the performance characteristics of a novel lightweight Al35Cr14Mg6Ti35V10 HEA, developed using SPS and mechanical alloying. This alloy was considered lightweight because it has a low density of 4.05 g cm^{-3}. The major elements having a purity of at least 99.95% were chosen due to their stability, low density, and compatibility with one another. The HEAs were synthesized by mechanical alloying in a Fritsch high-energy planetary ball mill, using tungsten carbide balls, and compacted via SPS. The alloying was achieved within 30 hours with a ball-to-power ratio of 10:1. The basic properties of the alloy studied were microstructural evolution, crystallographic phases, and hardness using SEM/EDS, XRD, and Vickers hardness tests, respectively. A single-phase BCC with slight impurities of tungsten carbide was obtained after 30 hours of milling. The final microstructures after SPS were two BCC phases and one HCP with a slightly embedded tungsten carbide contaminant. The hardness of Al35Cr14Mg6Ti35V10 HEA was established to be 460 HV, and this was adjudged superior to Mg, Ti, and Al-based conventional alloys. However, there was no recorded study of the corrosion properties of the material.

It has been identified from the literature reviewed that aluminium cookware is vulnerable to pitting corrosion, stress corrosion cracking, and leaching/wear.

Leaching was discovered to be predominantly common and higher with aluminium cookware used for cooking foods with relatively high acidic and salty content. Although findings showed that aluminium could protect itself through the formation of natural thin-film oxide, it erodes on continuous contact with saline or acidic medium. A good number of researchers agreed that the leaching of aluminium cookware is dangerous to human health and could also be harmful to the ecosystem. According to many authors, "over the years, anodization of aluminium cookware has been the major technique for improving and preserving its properties, however, it fails with time due to pitting and constant heating". Generally, authors agreed that HEAs have exceptional pitting corrosion resistance and outstanding mechanical properties, especially at elevated temperatures. Their properties were found comparably more favourable than most existing conventional high-temperature alloys and many superalloys. As mentioned earlier, studies have established that several HEAs have significantly improved strength-to-weight ratios, with a superior measure of tensile strength, fracture resistance, as well as oxidation and corrosion resistance than the usual alloys (Youssef et al., 2015; Yim et al., 2019; Ganesh & Raghavendra, 2020). The mechanical properties of any HEAs are a function of its crystal structure. BCC HEAs characteristically have low ductility and high YS, whereas FCC HEAs possess high ductility and low yield strength.

REFERENCES

Abdel-Gawad, S. A., Osman, W. M., & Fekry, A. M. (2019). Characterization and corrosion behavior of anodized aluminium alloys for military industries applications in artificial seawater. *Surfaces and Interfaces*, *14*, 314–323.

Akbar, S. I., & Nurpita, A. (2019). Performance analysis of aluminium industry based on product life cycle and market trends. *Review of Management and Entrepreneurship*, *3*(2), 91–106.

Akca, E., & Gursel, A. (2015). A review on superalloys and IN718 nickel-based Inconel superalloy. *Periodicals of Engineering and Natural Sciences*, *3*(1), 15–27.

Al Juhaiman, L. A. (2010). Estimating aluminium leaching from aluminium cook wares in different meat extracts and milk. *Journal of Saudi Chemical Society*, *14*(1), 131–137.

Al Juhaiman, L. A. (2016). Curcumin extract as a green inhibitor of leaching from aluminium cookware at quasi-cooking conditions. *Green and Sustainable Chemistry*, *6*(2), 57–70.

Al-Tamimi, W. H., & Mehdi, K. H. (2017). Inhibition of biogenic hydrogen sulfide produce by sulfate reducing bacteria isolated from oil fields in basra by nitrate-based treatment. *Journal of Petroleum Research & Studies*, *15*, 88–106.

Ayeni, O., Kambizi, L., Laubscher, C., Fatoki, O., & Olatunji, O. (2014). Risk assessment of wetland under aluminium and iron toxicities: A review. *Aquatic Ecosystem Health & Management*, *17*(2), 122–128.

Bała, P., Gorecki, K., Bednarczyk, W., Wątroba, M., Lech, S., & Kawałko, J. (2020). Effect of high-temperature exposure on the microstructure and mechanical properties of the $Al_5Ti_5Co_{35}Ni_{35}Fe_{20}$ high-entropy alloy. *Journal of Materials Research and Technology*, *9*(1), 551–559.

Belkhiri, L., Mouni, L., Narany, T. S., & Tiri, A. (2017). Evaluation of potential health risk of heavy metals in groundwater using the integration of indicator kriging and multivariate statistical methods. *Groundwater for Sustainable Development*, *4*, 12–22.

Bondy, S. C. (2016). Low levels of aluminium can lead to behavioral and morphological changes associated with Alzheimer's disease and age-related neurodegeneration. *Neurotoxicology*, *52*, 222–229.

Cao, Y., Liu, Y., Liu, B., & Zhang, W. (2018). Precipitation behavior during hot deformation of powder metallurgy Ti-Nb-Ta-Zr-Al high entropy alloys. *Intermetallics*, *100*, 95–103.

Chauhan, A., & Sharma, U. K. (2019). Influence of temperature and relative humidity variations on non-uniform corrosion of reinforced concrete. *Structures*, *19*, 296–308.

Chauhan, P., Yebaji, S., Nadakuduru, V. N., & Shanmugasundaram, T. (2020). Development of a novel light weight $Al_{35}Cr_{14}Mg_6Ti_{35}V_{10}$ high entropy alloy using mechanical alloying and spark plasma sintering. *Journal of Alloys and Compounds*, *820*, 153367.

Cheng, H., Liu, X., Tang, Q., Wang, W., Yan, X., & Dai, P. (2019). Microstructure and mechanical properties of FeCoCrNiMnAlx high-entropy alloys prepared by mechanical alloying and hot-pressed sintering. *Journal of Alloys and Compounds*, *775*, 742–751.

Choi, J., Seok, C. S., Park, S., & Kim, G. (2019). Effect of high-temperature degradation on microstructure evolution and mechanical properties of austenitic heat-resistant steel. *Journal of Materials Research and Technology*, *8*(2), 2011–2020.

Chuang, M. H., Tsai, M. H., Wang, W. R., Lin, S. J., & Yeh, J. W. (2011). Microstructure and wear behavior of $Al_xCo_{1.5}CrFeNi_{1.5}Ti_y$ high-entropy alloys. *Acta Materialia*, *59*(16), 6308–6317.

Cieslak, J., Tobola, J., Berent, K., & Marciszko, M. (2018). Phase composition of AlxFeNiCrCo high entropy alloys prepared by sintering and arc-melting methods. *Journal of Alloys and Compounds*, *740*, 264–272.

Diao, H., Santodonato, L. J., Tang, Z., Egami, T., & Liaw, P. K. (2015). Local structures of high-entropy alloys (HEAs) on atomic scales: An overview. *The Journal of the Minerals, Metals and Materials Society*, *67*(10), 2321–2325.

Divinski, S. V., Pokoev, A. V., Esakkiraja, N., & Paul, A. (2018). A mystery of" sluggish diffusion" in high-entropy alloys: The truth or a myth? *Diffusion Foundations*, *17*, 69–104.

Do Lee, C. (2018). Effect of artificial ageing on the defect susceptibility of tensile properties to porosity variation in A356 aluminium alloys. *International Journal of Metalcasting*, *12*(2), 321–330.

Dobbelstein, H., Gurevich, E. L., George, E. P., Ostendorf, A., & Laplanche, G. (2018). Laser metal deposition of a refractory TiZrNbHfTa high-entropy alloy. *Additive Manufacturing*, *24*, 386–390.

Elkilany, H. A., Shoeib, M. A., & Abdel-Salam, O. E. (2019). Influence of hard anodizing on the mechanical and corrosion properties of different aluminium alloys. *Metallography, Microstructure, and Analysis*, *8*(6), 861–870.

Erdogan, A., Yener, T., & Zeytin, S. (2018). Fast production of high entropy alloys (CoCrFeNiAlxTiy) by electric current activated sintering system. *Vacuum*, *155*, 64–72.

Fang, Z., Cao, J., & Guan, Y. (2020a). Anticorrosion structural design for aluminium alloy vessel. *Corrosion Control Technologies for Aluminium Alloy Vessel*, *1*, 343–357.

Fang, Z., Cao, J., & Guan, Y. (2020b). Maintenance and repair of aluminium alloy vessel anticorrosion system. *Corrosion Control Technologies for Aluminium Alloy Vessel*, *10*, 77–438.

Fayomi, O. S. I., Akande, I. G., Popoola, A. P. I., & Molifi, H. (2019). Potentiodynamic polarization studies of Cefadroxil and Dicloxacillin drugs on the corrosion susceptibility of aluminium AA6063 in 0.5 M nitric acid. *Journal of Materials Research and Technology*, *8*(3), 3088–3096.

Feuerbacher, M., Lienig, T., & Thomas, C. (2018). A single-phase bcc high-entropy alloy in the refractory Zr-Nb-Ti-V-Hf system. *Scripta Materialia*, *152*, 40–43.

Fu, X., Schuh, C. A., & Olivetti, E. A. (2017). Materials selection considerations for high entropy alloys. *Scripta Materialia*, *138*, 145–150.

Ganesh, U. L., & Raghavendra, H. (2020). Review on the transition from conventional to multi-component-based nano-high-entropy alloys—NHEAs. *Journal of Thermal Analysis and Calorimetry*, *139*(1), 207–216.

Gao, M. C., Zhang, B., Guo, S. M., Qiao, J. W., & Hawk, J. A. (2016). High-entropy alloys in hexagonal close-packed structure. *Metallurgical and Materials Transactions A*, *47*(7), 3322–3332.

Gatti, A. M., & Montanari, S. (2018). Food contamination: From food degradation to food-borne diseases. In *Food safety and preservation* (pp. 431–456). Academic Press.

George, E. P., Raabe, D., & Ritchie, R. O. (2019). High-entropy alloys. *Nature Reviews Materials*, *4*(8), 515–534.

Ghasemidehkordi, B., Nazem, H., Malekirad, A. A., Fazilati, M., Salavati, H., & Rezaei, M. (2018). Human health risk assessment of aluminium via consumption of contaminated vegetables. *Quality Assurance and Safety of Crops & Foods*, *10*(2), 115–123.

Goebel, J., Ghidini, T., & Graham, A. J. (2016). Stress-corrosion cracking characterisation of the advanced aerospace Al–Li 2099-T86 alloy. *Materials Science and Engineering: A*, *673*, 16–23.

Goswami, S. P., Maurya, B. R., Dubey, A. N., & Singh, N. K. (2019). Role of phosphorus solubilizing microorganisms and dissolution of insoluble phosphorus in soil. *International Journal of Chemical Studies*, *7*(3), 3905–3913.

Goyal, A., Singh, R., & Singh, G. (2017). Study of high-temperature corrosion behavior of D-gun spray coatings on ASTM-SA213, T-11 steel in molten salt environment. *Materials Today: Proceedings*, *4*(2), 142–151.

Grgur, B., & Marunkić, L. (2018). The influence of chloride anions on the pitting corrosion of aluminum alloy en 46000. *Zaštita materijala*, *59*(2), 243–248.

Gu, J., Ding, J., Williams, S. W., Gu, H., Bai, J., Zhai, Y., & Ma, P. (2016). The strengthening effect of inter-layer cold working and post-deposition heat treatment on the additively manufactured Al–63 Cu alloy. *Materials Science and Engineering: A*, *651*, 18–26.

Gu, T., Jia, R., Unsal, T., & Xu, D. (2019). Toward a better understanding of microbiologically influenced corrosion caused by sulfate reducing bacteria. *Journal of Materials Science & Technology*, *35*(4), 631–636.

Guan, F., Zhai, X., Duan, J., Zhang, J., Li, K., & Hou, B. (2017). Influence of sulfate-reducing bacteria on the corrosion behavior of 5052 aluminium alloy. *Surface and Coatings Technology*, *316*, 171–179.

Guo, S. (2015). Phase selection rules for cast high entropy alloys: An overview. *Materials Science and Technology, 31*(10), 1223–1230.

Guo, W., Liu, B., Liu, Y., Li, T., Fu, A., Fang, Q., & Nie, Y. (2019). Microstructures and mechanical properties of ductile NbTaTiV refractory high entropy alloy prepared by powder metallurgy. *Journal of Alloys and Compounds*, *776*, 428–436.

Gupta, Y. K., Meenu, M., & Peshin, S. S. (2019). Aluminium utensils: Is it a concern? *The National Medical Journal of India*, *32*(1), 38–40.

Gwalani, B., Soni, V., Waseem, O. A., Mantri, S. A., & Banerjee, R. (2019). Laser additive manufacturing of compositionally graded AlCrFeMoVx (x= 0 to 1) high-entropy alloy system. *Optics & Laser Technology*, *113*, 330–337.

Hache, M. J., Cheng, C., & Zou, Y. (2020). Nanostructured high-entropy materials. *Journal of Materials Research*, *35*(8), 1051–1075.

Hallensleben, P., Schaar, H., Thome, P., Jons, N., Jafarizadeh, A., Steinbach, I., ... & Frenzel, J. (2017). On the evolution of cast microstructures during processing of single crystal Ni-base superalloys using a Bridgman seed technique. *Materials & Design*, *128*, 98–111.

Hemphill, M. A., Yuan, T., Wang, G. Y., Yeh, J. W., Tsai, C. W., Chuang, A., & Liaw, P. K. (2012). Fatigue behavior of $Al_{0.5}$CoCrCuFeNi high entropy alloys. *Acta Materialia*, *60*(16), 5723–5734.

Hou, B., Li, X., Ma, X., Du, C., Zhang, D., Zheng, M., & Ma, F. (2017). The cost of corrosion in China. *Npj Materials Degradation*, *1*(1), 1–10.

Hou, L., Hui, J., Yao, Y., Chen, J., & Liu, J. (2019). Effects of boron content on microstructure and mechanical properties of AlFeCoNiBx high entropy alloy prepared by vacuum arc melting. *Vacuum*, *164*, 212–218.

Hu, P., Meng, Q., Hu, W., Shen, F., Zhan, Z., & Sun, L. (2016). A continuum damage mechanics approach coupled with an improved pit evolution model for the corrosion fatigue of aluminium alloy. *Corrosion Science*, *113*, 78–90.

Huang, I. W., Hurley, B. L., Yang, F., & Buchheit, R. G. (2016). Dependence on temperature, pH, and Cl^- in the uniform corrosion of aluminium alloys 2024-T3, 6061-T6, and 7075-T6. *Electrochimica Acta*, *199*, 242–253.

Ilich, J. Z., & Kerstetter, J. E. (2000). Nutrition in bone health revisited: A story beyond calcium. *Journal of the American College of Nutrition*, *19*(6), 715–737.

Iqbal, H., Tesfamariam, S., Haider, H., & Sadiq, R. (2017). Inspection and maintenance of oil & gas pipelines: a review of policies. *Structure and Infrastructure Engineering*, *13*(6), 794–815.

Jabeen, S., Ali, B., Khan, M. A., Khan, M. B., & Hasan, S. A. (2016). Aluminium intoxication through leaching in food preparation. *Alexandria Science Exchange Journal*, *37*(4), 618–626.

Jariyaboon, M., Moller, P., & Ambat, R. (2012). Effect of pressurized steam on AA1050 aluminium. *Anti-Corrosion Methods and Materials*, *59*(3), 103–109.

Jekle, M., Horeld, C., Gratzl, R., Roth, M., Becker, T., &Hobel, W. (2016). Aluminium leaching from baking tray materials into surface-alkalized baked products. *Cereal Technology*, *3*, 127–135.

Jennrich, P., & Schulte-Uebbing, C. (2016). Does aluminium trigger breast cancer? *The Open Access Journal of Science and Technology*, *4*(3), 1–6.

Ji, W., Wang, W., Wang, H., Zhang, J., Wang, Y., Zhang, F., & Fu, Z. (2015). Alloying behavior and novel properties of CoCrFeNiMn high-entropy alloy fabricated by mechanical alloying and spark plasma sintering. *Intermetallics*, *56*, 24–27.

Jiang, Y. Q., Li, J., Juan, Y. F., Lu, Z. J., & Jia, W. L. (2019). Evolution in microstructure and corrosion behavior of AlCoCrxFeNi high-entropy alloy coatings fabricated by laser cladding. *Journal of Alloys and Compounds*, *775*, 1–14.

Joo, S. H., Kato, H., Jang, M. J., Moon, J., Kim, E. B., Hong, S. J., & Kim, H. S. (2017). Structure and properties of ultrafine-grained CoCrFeMnNi high-entropy alloys produced by mechanical alloying and spark plasma sintering. *Journal of Alloys and Compounds*, *698*, 591–604.

Joseph, J., Haghdadi, N., Shamlaye, K., Hodgson, P., Barnett, M., & Fabijanic, D. (2019). The sliding wear behaviour of CoCrFeMnNi and AlxCoCrFeNi high entropy alloys at elevated temperatures. *Wear*, *428*, 32–44.

Juan, C. C., Tsai, M. H., Tsai, C. W., Lin, C. M., Wang, W. R., Yang, C. C., &Yeh, J. W. (2015). Enhanced mechanical properties of HfMoTaTiZr and HfMoNbTaTiZr refractory high-entropy alloys. *Intermetallics*, *62*, 76–83.

Kandimalla, R., Vallamkondu, J., Corgiat, E. B., & Gill, K. D. (2016). Understanding aspects of aluminium exposure in Alzheimer's disease development. *Brain Pathology*, *26*(2), 139–154.

Karlsson, D., Marshal, A., Johansson, F., Schuisky, M., Sahlberg, M., Schneider, J. M., & Jansson, U. (2019). Elemental segregation in an AlCoCrFeNi high-entropy alloy – A comparison between selective laser melting and induction melting. *Journal of Alloys and Compounds*, *784*, 195–203.

Kim, B., Ryu, C., Lee, U., Kim, Y., Lee, J., & Song, J. (2017). A technical review on the protective measures of high temperature corrosion of boiler heat exchangers with additives. *Clean Technology*, *23*(3), 223–236.

Kim, J., Wakai, A., & Moridi, A. (2020). Materials and manufacturing renaissance: Additive manufacturing of high-entropy alloys. *Journal of Materials Research*, *35*(15), 1963–1983.

Klein, G. L. (2019). Aluminium toxicity to bone: A multisystem effect? *Osteoporosis and Sarcopenia*, *5*(1), 2–5.

Klotz, K., Weistenhofer, W., Neff, F., Hartwig, A., Van-Thriel, C., & Drexler, H. (2017). The health effects of aluminium exposure. *DeutschesArzteblatt International*, *114*(39), 653–659.

Koch, G., Varney, J., Thompson, N., Moghissi, O., Gould, M., & Payer, J. (2016). International measures of prevention, application, and economics of corrosion technologies study. *NACE International*, 216 pages.

Koli, D. K., Agnihotri, G., & Purohit, R. (2015). Advanced aluminium matrix composites: The critical need of automotive and aerospace engineering fields. *Materials Today: Proceedings*, *2*(4-5), 3032–3041.

Kumar, A., Swarnakar, A. K., & Chopkar, M. (2018). Phase evolution and mechanical properties of AlCoCrFeNiSi x high-entropy alloys synthesized by mechanical alloying and spark plasma sintering. *Journal of Materials Engineering and Performance*, *27*(7), 3304–3314.

Kumar, D., Maulik, O., Kumar, S., Prasad, Y. V. S. S., & Kumar, V. (2018). Phase and thermal study of equiatomic AlCuCrFeMnW high entropy alloy processed via spark plasma sintering. *Materials Chemistry and Physics*, *210*, 71–77.

Laplanche, G., Berglund, S., Reinhart, C., Kostka, A., Fox, F., & George, E. P. (2018). Phase stability and kinetics of σ-phase precipitation in CrMnFeCoNi high-entropy alloys. *Acta Materialia*, *161*, 338–351.

Laurent-Brocq, M., Akhatova, A., Perriere, L., Chebini, S., Sauvage, X., Leroy, E., & Champion, Y. (2015). Insights into the phase diagram of the CrMnFeCoNi high entropy alloy. *Acta Materialia*, *88*, 355–365.

Li, D., & Zhang, Y. (2016). The ultrahigh Charpy impact toughness of forged AlxCoCrFeNi high entropy alloys at room and cryogenic temperatures. *Intermetallics*, *70*, 24–28.

Li, W., Liu, P., & Liaw, P. K. (2018). Microstructures and properties of high-entropy alloy films and coatings: A review. *Materials Research Letters*, *6*(4), 199–229.

Li, Z., Lv, S. L., Liu, Z. E., & Zhang, W. (2018). Effect of surface quality on pitting corrosion behavior of aluminium alloy 2A12. *Proceedings of the International Workshop on Materials, Chemistry and Engineering*, *1*, 324–329.

Li, Z., Pradeep, K. G., Deng, Y., Raabe, D., & Tasan, C. C. (2016). Metastable high-entropy dual-phase alloys overcome the strength–ductility trade-off. *Nature*, *534*(7606), 227–230.

Liang, W., Qinglin, P. A. N., Yunbin, H. E., Yunchun, L. I., Yingchun, Z. H. O. U., & Congge, L. U. (2008). Effect of aging on the mechanical properties and corrosion susceptibility of an Al-Cu-Li-Zr alloy containing Sc. *Rare Metals*, *27*(2), 146–152.

Liu, W. H., He, J. Y., Huang, H. L., Wang, H., Lu, Z. P., & Liu, C. T. (2015). Effects of Nb additions on the microstructure and mechanical property of CoCrFeNi high-entropy alloys. *Intermetallics*, *60*, 1–8.

Lu, Y., Dong, Y., Jiang, H., Wang, Z., Cao, Z., Guo, S., ... & Liaw, P. K. (2020). Promising properties and future trend of eutectic high entropy alloys. *Scripta Materialia*, *187*, 202–209.

Lyu, Z., Fan, X., Lee, C., Wang, S. Y., Feng, R., & Liaw, P. K. (2018). Fundamental understanding of mechanical behavior of high-entropy alloys at low temperatures: A review. *Journal of Materials Research, 33*(19), 2998–3010.

Lyu, Z., Lee, C., Wang, S. Y., Fan, X., Yeh, J. W., & Liaw, P. K. (2019). Effects of constituent elements and fabrication methods on mechanical behavior of high-entropy alloys: A review. *Metallurgical and Materials Transactions A, 50*(1), 1–28.

Macdonald, B. E., Fu, Z., Zheng, B., Chen, W., Lin, Y., Chen, F., & Lavernia, E. J. (2017). Recent progress in high entropy alloy research. *Journal of Metals, 69*(10), 2024–2031.

Mai, W., & Soghrati, S. (2017). A phase field model for simulating the stress corrosion cracking initiated from pits. *Corrosion Science, 125*, 87–98.

Mannava, V., SambasivaRao, A., Kamaraj, M., & Kottada, R. S. (2019). Influence of two different salt mixture combinations of Na_2SO_4-NaCl-$NaVO_3$ on hot corrosion behavior of Ni-base superalloy Nimonic 263 at 800° C. *Journal of Materials Engineering and Performance, 28*(2), 1077–1093.

Martínez, J., Villacís, S. P., Orozco, M. A., & Vaca, D. E. (2015).Corrosion analysis in different materials for induction cookware. *In Congreso de Ciencia y Tecnología ESPE, 10*(1), 155–159.

Martinez-Salazar, A. L., Melo-Banda, J. A., Coronel-García, M. A., González-Barbosa, J. J., & Domínguez-Esquivel, J. M. (2020). Hydrogen generation by aluminium alloy corrosion in aqueous acid solutions promoted by nanometal: Kinetics study. *Renewable Energy, 146*, 2517–2523.

Melchers, R. E. (2020). A review of trends for corrosion loss and pit depth in longer-term exposures. *Corrosion and Materials Degradation, 1*(1), 42–58.

Miracle, D. B., & Senkov, O. N. (2017). A critical review of high entropy alloys and related concepts. *Acta Materialia, 122*, 448–511.

Miroshnikova, T., & Kuchugin, N. (2015). Economic efficiency of innovative corrosion resistant coating for sectors of Russian national economy. *Procedia Economics and Finance, 24*, 426–434.

Mol, S., & Ulusoy, S. (2020). The effect of cooking conditions on aluminum concentrations of seafood, cooked in aluminum foil. *Journal of Aquatic Food Product Technology, 29*(2), 186–193.

Murali, M., Babu, S. K., Krishna, B. J., & Vallimanalan, A. (2016). Synthesis and characterization of AlCoCrCuFeZnx high-entropy alloy by mechanical alloying. *Progress in Natural Science: Materials International, 26*(4), 380–384.

Nathan, N., Sileo, C., Calender, A., Pacheco, Y., Rosental, P. A., Cavalin, C., & Silicosis research group (2019). Paediatric sarcoidosis. *Paediatric Respiratory Reviews, 29*, 53–59.

Nong, Z. S., Lei, Y. N., & Zhu, J. C. (2018). Wear and oxidation resistances of AlCrFeNiTi-based high entropy alloys. *Intermetallics, 101*, 144–151.

Odularu, A. T., Ajibade, P. A., & Onianwa, P. C. (2013). Comparative study of leaching of aluminium from aluminium, clay, stainless steel, and steel cooking pots. *ISRN Public Health, 2013*, 1–4.

Orisanmi, B. O., Afolalu, S. A., Adetunji, O. R., Salawu, E. Y., Okokpujie, I. P., Abioye, A. A., ... & Abioye, O. P. (2017). Cost of corrosion of metallic products in Federal University of Agriculture, Abeokuta. *International Journal of Applied Engineering, 12*(24), 14141–14147.

Ossai, C. I., Boswell, B., & Davies, I. J. (2015)., C. I., Boswell, B., & Davies, I. J. (2015). Pipeline failures in corrosive environments – A conceptual analysis of trends and effects. *Engineering Failure Analysis, 53*, 36–58.

Otto, F., Dlouh, A., Somsen, C., Bei, H., Eggeler, G., & George, E. P. (2013). The influences of temperature and microstructure on the tensile properties of a CoCrFeMnNi high-entropy alloy. *Acta Materialia, 61*(15), 5743–5755.

Pan, J., Dai, T., Lu, T., Ni, X., Dai, J., & Li, M. (2018). Microstructure and mechanical properties of $Nb_{25}Mo_{25}Ta_{25}W_{25}$ and $Ti_8Nb_{23}Mo_{23}Ta_{23}W_{23}$ high entropy alloys prepared by mechanical alloying and spark plasma sintering. *Materials Science and Engineering: A, 738*, 362–366.

Park, H. J., Na, Y. S., Hong, S. H., Kim, J. T., Kim, Y. S., Lim, K. R., ... & Kim, K. B. (2016). Phase evolution, microstructure and mechanical properties of equi-atomic substituted TiZrHfNiCu and TiZrHfNiCuM (M= Co, Nb) high-entropy alloys. *Metals and Materials International, 22*(4), 551–556.

Patil, M. D. D., & Ghatge, D. A. (2017). Parametric evaluation of melting practice on induction furnace to improve efficiency and system productivity of CI and SGI foundry-A review. *International Advanced Research Journal in Science, Engineering & Technology, 4*, 159–163.

Pickering, E. J., & Jones, N. G. (2016). High-entropy alloys: A critical assessment of their founding principles and future prospects. *International Materials Reviews, 61*(3), 183–202.

Poulia, A., Georgatis, E., & Karantzalis, A. (2019). Evaluation of the microstructural aspects, mechanical properties and dry sliding wear response of MoTaNbVTi refractory high entropy alloy. *Metals and Materials International, 25*(6), 1529–1540.

Prabhu, T. R., Arivarasu, M., Chodancar, Y., Arivazhagan, N., Sumanth, G., & Mishra, R. K. (2019). Tribological behaviour of graphite-reinforced FeNiCrCuMo high-entropy alloy self-lubricating composites for aircraft braking energy applications. *Tribology Letters, 67*(3), 78–93.

Qiu, Y., Gibson, M. A., Fraser, H. L., & Birbilis, N. (2015). Corrosion characteristics of high entropy alloys. *Materials Science and Technology, 31*(10), 1235–1243.

Roshani, M., Rouhaghdam, A. S., Aliofkhazraei, M., & Astaraee, A. H. (2017). Optimization of mechanical properties for pulsed anodizing of aluminium. *Surface and Coatings Technology, 310*, 17–24.

Santodonato, L. J., Zhang, Y., Feygenson, M., Parish, C. M., Gao, M. C., Weber, R. J., ... & Liaw, P. K. (2015). Deviation from high-entropy configurations in the atomic distributions of a multi-principal-element alloy. *Nature Communications, 6*(1), 1–13.

Sarswat, P. K., Sarkar, S., Murali, A., Huang, W., Tan, W., & Free, M. L. (2019). Additive manufactured new hybrid high entropy alloys derived from the AlCoFeNiSmTiVZr system. *Applied Surface Science, 476*, 242–258.

Seifi, M., Li, D., Yong, Z., Liaw, P. K., & Lewandowski, J. J. (2015). Fracture toughness and fatigue crack growth behavior of as-cast high-entropy alloys. *The Journal of the Minerals, Metals and Materials Society, 67*(10), 2288–2295.

Sekhar, R. A., Samal, S., Nayan, N., & Bakshi, S. R. (2019). Microstructure and mechanical properties of Ti-Al-Ni-Co-Fe based high entropy alloys prepared by powder metallurgy route. *Journal of Alloys and Compounds, 787*, 123–132.

Senkov, O. N., & Miracle, D. B. (2016). A new thermodynamic parameter to predict formation of solid solution or intermetallic phases in high entropy alloys. *Journal of Alloys and Compounds, 658*, 603–607.

Senkov, O. N., Wilks, G. B., Scott, J. M., & Miracle, D. B. (2011). Mechanical properties of $Nb_{25}Mo_{25}Ta_{25}W_{25}$ and $V_{20}Nb_{20}Mo_{20}Ta_{20}W_{20}$ refractory high entropy alloys. *Intermetallics, 19*(5), 698–706.

Shang, X., Wang, Z., He, F., Wang, J., Li, J., & Yu, J. (2018). The intrinsic mechanism of corrosion resistance for FCC high entropy alloys. *Science China Technological Sciences, 61*(2), 189–196.

Shi, P., & Mahadevan, S. (2001). Damage tolerance approach for probabilistic pitting corrosion fatigue life prediction. *Engineering Fracture Mechanics, 68*(13), 1493–1507.

Shi, P., Ren, W., Zheng, T., Ren, Z., Hou, X., Peng, J., ... & Liaw, P. K. (2019). Enhanced strength–ductility synergy in ultrafine-grained eutectic high-entropy alloys by inheriting microstructural lamellae. *Nature Communications*, *10*(1), 489.

Sielski, R. A. (2008). Research needs in aluminium structure. *Ships and Offshore Structures*, *3*(1), 57–65.

Sim, H. S., Plascencia, M. A., Vargas, A., & Karagozian, A. R. (2020). Acoustically forced droplet combustion of liquid fuel with reactive aluminium nanoparticulates. *Combustion Science and Technology*, *192*(5), 761–785.

Skoczko, I., & Szatyłowicz, E. (2019). Treatment method assessment of the impact on the corrosivity and aggressiveness for the boiler feed water. *Water*, *11*(10), 1–14.

Soares, C. G., Garbatov, Y., Zayed, A., & Wang, G. (2009). Influence of environmental factors on corrosion of ship structures in marine atmosphere. *Corrosion Science*, *51*(9), 2014–2026.

Soares, D. S., Bolgar, G., Dantas, S. T., Augusto, P. E., & Soares, B. M. (2019). Interaction between aluminium cans and beverages: Influence of catalytic ions, alloy and coating in the corrosion process. *Food Packaging and Shelf Life*, *19*, 56–65.

Sriraman, M. R., & Pidaparti, R. M. (2010). Crack initiation life of materials under combined pitting corrosion and cyclic loading. *Journal of Materials Engineering and Performance*, *19*(1), 7–12.

Stahl, T., Falk, S., Taschan, H., Boschek, B., & Brunn, H. (2018). Evaluation of human exposure to aluminium from food and food contact materials. *European Food Research and Technology*, *244*(12), 2077–2084.

Staley, J. T. (2017). Corrosion of aluminium aerospace alloys. *Materials Science*, *877*, 485–491.

Stemper, L., Mitas, B., Kremmer, T., Otterbach, S., Uggowitzer, P. J., & Pogatscher, S. (2019). Age-hardening of high pressure die casting AlMg alloys with Zn and combined Zn and Cu additions. *Materials and Design*, *181*, 1–11.

Takeuchi, A., Amiya, K., & Yubuta, K. (2018). Partially-devitrified icosahedral quasicrystalline phase in $Ti_{33\cdot33}Zr_{33\cdot33}Hf_{13\cdot33}Ni_{20}$ and $Zr_{30}Hf_{30}Ni_{15}Cu_{10}Ti_{15}$ amorphous alloys with near equi-atomic compositions. *Materials Chemistry and Physics*, *210*, 245–250.

Thompson, N. G., Yunovich, M., & Dunmire, D. (2007). Cost of corrosion and corrosion maintenance strategies. *Corrosion Reviews*, *25*(3-4), 247–262.

Tian, Y., Lu, C., Shen, Y., & Feng, X. (2019). Microstructure and corrosion property of CrMnFeCoNi high entropy alloy coating on Q235 substrate via mechanical alloying method. *Surfaces and Interfaces*, *15*, 135–140.

Tisza, M., & Czinege, I. (2018). Comparative study of the application of steels and aluminium in light-weight production of automotive parts. *International Journal of Light-Weight Materials and Manufacture*, *1*(4), 229–238.

Torralba, J. M., Alvaredo, P., & García-Junceda, A. (2019). High-entropy alloys fabricated via powder metallurgy. A critical review. *Powder Metallurgy*, *62*(2), 84–114.

Tsau, C. H., & Wang, W. L. (2018). Microstructures, hardness and corrosion behaviors of $FeCoNiNb_{0.5}Mo_{0.5}$ and FeCoNiNb high-entropy alloys. *Materials*, *11*(1), 1–12.

Tseng, K., Yang, Y., Juan, C., Chin, T., Tsai, C., & Yeh, J. (2018). A light-weight high-entropy alloy $Al_{20}Be_{20}Fe_{10}Si_{15}Ti_{35}$. *Science China Technological Sciences*, *61*(2), 184–188.

Vaidya, M., Anupam, A., Bharadwaj, J. V., Srivastava, C., & Murty, B. S. (2019). Grain growth kinetics in CoCrFeNi and CoCrFeMnNi high entropy alloys processed by spark plasma sintering. *Journal of Alloys and Compounds*, *791*, 1114–1121.

Vaidya, M., Muralikrishna, G. M., & Murty, B. S. (2019). High-entropy alloys by mechanical alloying: A review. *Journal of Materials Research*, *34*(5), 664–686.

Vu, C., Wern, C., Kim, B. H., Kim, S. K., Choi, H. J., & Yi, S. (2019). Fatigue characteristic analysis of new ECO7175v1 extruded aluminium alloy. *Journal of Aerospace Engineering*, *32*(1), 1–6.

Wang, M., Ma, Z. L., Xu, Z. Q., & Cheng, X. W. (2021). Designing VxNbMoTa refractory high-entropy alloys with improved properties for high-temperature applications. *Scripta Materialia*, *191*, 131–136.

Wang, R. N., Xu, Q. Y., & Liu, B. C. (2018). A model for simulation of recrystallization microstructure in single-crystal superalloy. *Rare Metals*, *37*(12), 1027–1034.

Wang, W. R., Wang, W. L., & Yeh, J. W. (2014). Phases, microstructure and mechanical properties of AlxCoCrFeNi high-entropy alloys at elevated temperatures. *Journal of Alloys and Compounds*, *589*, 143–152.

Wang, W., Duong-Viet, C., Xu, Z., Ba, H., Tuci, G., Giambastiani, G., ... & Pham-Huu, C. (2020). CO_2 methanation under dynamic operational mode using nickel nanoparticles decorated carbon felt (Ni/OCF) combined with inductive heating. *Catalysis Today*, *357*, 214–220.

Wang, W., Li, B., Zhai, S., Xu, J., Niu, Z., Xu, J., & Wang, Y. (2018). Alloying behavior and properties of FeSiBAlNiCo x high entropy alloys fabricated by mechanical alloying and spark plasma sintering. *Metals and Materials International*, *24*(5), 1112–1119.

Waseem, O. A., Lee, J., Lee, H. M., & Ryu, H. J. (2018). The effect of Ti on the sintering and mechanical properties of refractory high-entropy alloy TixWTaVCr fabricated via spark plasma sintering for fusion plasma-facing materials. *Materials Chemistry and Physics*, *210*, 87–94.

Wei, R., Sun, H., Chen, C., Tao, J., & Li, F. (2018). Formation of soft magnetic high entropy amorphous alloys composites containing in situ solid solution phase. *Journal of Magnetism and Magnetic Materials*, *449*, 63–67.

Weidenhamer, J. D., Fitzpatrick, M. P., Biro, A. M., Kobunski, P. A., Hudson, M. R., Corbin, R. W., & Gottesfeld, P. (2017). Metal exposures from aluminium cookware: An unrecognized public health risk in developing countries. *Science of the Total Environment*, *579*, 805–813.

Wu, Z., Bei, H., Otto, F., Pharr, G. M., & George, E. P. (2014). Recovery, recrystallization, grain growth and phase stability of a family of FCC-structured multi-component equiatomic solid solution alloys. *Intermetallics*, *46*, 131–140.

Xiao, D. H., Zhou, P. F., Wu, Z., Bei, H., Otto, F., Pharr, G. M., & George, E. P. (2014). Microstructure, mechanical and corrosion behaviours of AlCoCuFeNi-(Cr, Ti) high entropy alloys. *Materials and Design*, *116*, 438–447.

Xie, Y., Cheng, H., Tang, Q., Chen, W., Chen, W., & Dai, P. (2018). Effects of N addition on microstructure and mechanical properties of CoCrFeNiMn high entropy alloy produced by mechanical alloying and vacuum hot pressing sintering. *Intermetallics*, *93*, 228–234.

Yang, L., Li, S., Chang, X., Zhong, H., & Fu, H. (2015). Twinned dendrite growth during Bridgman solidification. *Acta Materialia*, *97*, 269–281.

Ye, Y. F., Wang, Q., Lu, J., Liu, C. T., & Yang, Y. (2016). High-entropy alloy: Challenges and prospects. *Materials Today*, *19*(6), 349–362.

Yim, D., Sathiyamoorthi, P., Hong, S. J., & Kim, H. S. (2019). Fabrication and mechanical properties of TiC reinforced CoCrFeMnNi high-entropy alloy composite by water atomization and spark plasma sintering. *Journal of Alloys and Compounds*, *781*, 389–396.

Yin, Y., Zhang, J., Tan, Q., Zhuang, W., Mo, N., Bermingham, M., & Zhang, M. X. (2019). Novel cost-effective Fe-based high entropy alloys with balanced strength and ductility. *Materials and Design*, *162*, 24–33.

Youssef, K. M., Zaddach, A. J., Scattergood, R. O., & Koch, C. C. (2015). A novel low-density, high-hardness, high-entropy alloy with close-packed single-phase nanocrystalline structures. *Materials Research Letters*, *3*(2), 95–99.

Yu, P. F., Zhang, L. J., Cheng, H., Zhang, H., Ma, M. Z., Li, Y. C., ... & Liu, R. P. (2016). The high-entropy alloys with high hardness and soft magnetic property prepared by mechanical alloying and high-pressure sintering. *Intermetallics*, *70*, 82–87.

Zaddach, A. J., Scattergood, R. O., & Koch, C. C. (2015). Tensile properties of low-stacking fault energy high-entropy alloys. *Materials Science and Engineering: A*, *636*, 373–378.

Zhang, M., Zhang, W., Liu, Y., Liu, B., & Wang, J. (2018). FeCoCrNiMo high-entropy alloys prepared by powder metallurgy processing for diamond tool applications. *Powder Metallurgy*, *61*(2), 123–130.

Zhang, P., Li, Y., Chen, Z., Zhang, J., & Shen, B. (2019). Oxidation response of a vacuum arc melted NbZrTiCrAl refractory high entropy alloy at 800–1200°C. *Vacuum*, *162*, 20–27.

Zhang, W., Liaw, P. K., & Zhang, Y. (2018). Science and technology in high-entropy alloys. *Science China Materials*, *61*(1), 2–22.

Zhang, Y., Zuo, T. T., Tang, Z., Gao, M. C., Dahmen, K. A., Liaw, P. K., & Lu, Z. P. (2014). Microstructures and properties of high-entropy alloys. *Progress in Materials Science*, *61*, 1–93.

Zhao, D., Zhuang, Z., Cao, X., Zhang, C., Peng, Q., Chen, C., & Li, Y. (2020). Atomic site electro catalysts for water splitting, oxygen reduction and selective oxidation. *Chemical Society Reviews*, *49*(7), 2215–2264.

Zou, Y., Maiti, S., Steurer, W., & Spolenak, R. (2014). Size-dependent plasticity in a $Nb_{25}Mo_{25}Ta_{25}W_{25}$ refractory high-entropy alloy. *Acta Materialia*, *65*, 85–97.

Index

For Product Safety Concerns and Information please contact our EU representative GPSR@taylorandfrancis.com Taylor & Francis Verlag GmbH, Kaufingerstraße 24, 80331 München, Germany

Batch number: 10397790

Printed by Printforce, the Netherlands